Low-Power Circuits for Emerging Applications in Communications, Computing, and Sensing

Devices, Circuits, and Systems

Series Editor

Krzysztof Iniewski
Emerging Technologies CMOS Inc.
Vancouver, British Columbia, Canada

PUBLISHED TITLES:

3D Integration in VLSI Circuits
Implementation Technologies and Applications
Katsuyuki Sakuma

Advances in Imaging and Sensing
Shuo Tang and Daryoosh Saeedkia

Analog Electronics for Radiation Detection
Renato Turchetta

Atomic Nanoscale Technology in the Nuclear Industry
Taeho Woo

Biological and Medical Sensor Technologies
Krzysztof Iniewski

Biomaterials and Immune Response
Complications, Mechanisms and Immunomodulation
Nihal Engin Vrana

Building Sensor Networks
From Design to Applications
Ioanis Nikolaidis and Krzysztof Iniewski

Cell and Material Interface
Advances in Tissue Engineering, Biosensor, Implant, and Imaging
Technologies
Nihal Engin Vrana

Circuits and Systems for Security and Privacy
Farhana Sheikh and Leonel Sousa

Circuits at the Nanoscale
Communications, Imaging, and Sensing
Krzysztof Iniewski

CMOS
Front-End Electronics for Radiation Sensors
Angelo Rivetti

CMOS Time-Mode Circuits and Systems
Fundamentals and Applications
Fei Yuan

Low-Power Circuits for Emerging Applications in Communications, Computing, and Sensing

Edited by
Fei Yuan

Managing Editor
Krzysztof Iniewski

CRC Press
Taylor & Francis Group
Boca Raton London New York

CRC Press is an imprint of the
Taylor & Francis Group, an **informa** business

CRC Press
Taylor & Francis Group
6000 Broken Sound Parkway NW, Suite 300
Boca Raton, FL 33487-2742

First issued in paperback 2020

© 2019 by Taylor & Francis Group, LLC
CRC Press is an imprint of Taylor & Francis Group, an Informa business

No claim to original U.S. Government works

ISBN-13: 978-1-138-58001-5 (hbk)
ISBN-13: 978-0-367-73214-1 (pbk)

Library of Congress Cataloging-in-Publication Data

Names: Iniewski, Krzysztof, 1960- editor. | Yuan, Fei, 1964- editor.
Title: Low power circuits for emerging applications in communications, computing, and sensing / Krzysztof Iniewski and Fei Yuan, editors.
Description: First edition. | Boca Raton : CRC Press / Taylor & Francis, [2018] | Series: Taylor and Francis series in devices, circuits, & systems | Includes bibliographical references.
Identifiers: LCCN 2018033078 | ISBN 9781138580015 (hardback : alk. paper) | ISBN 9780429507564 (ebook)
Subjects: LCSH: Low voltage integrated circuits. | Digital electronics.
Classification: LCC TK7874.66 .L6486 2018 | DDC 621.381/32-dc23
LC record available at https://lccn.loc.gov/2018033078

Visit the Taylor & Francis Web site at
http://www.taylorandfrancis.com

and the CRC Press Web site at
http://www.crcpress.com

Contents

Preface

Making processing information more energy-efficient would save money, reduce energy use, and permit batteries that provide power in mobile devices to run longer or be smaller. New approaches to lower energy requirements in computing, communication, and sensing need to be investigated. This book addresses this need in multiple application areas and serves as a guide into emerging circuit technologies.

Revolutionary device concepts, sensors, and associated circuits and architectures that will greatly extend the practical engineering limits of energy-efficient computation are being investigated. Disruptive new system architectures, circuit microarchitectures, and attendant device and interconnect technology aimed at achieving the highest level of computational energy efficiency for general-purpose computing systems need to be developed. This book provides chapters dedicated to such efforts from the device to the circuit to the system level.

Contributors

Fady Abouzeid
STMicroelectronics
Crolles, France

Pao-Lung Chen
Kairos Microsystems Corporation
Melrose, Florida

Martin Cochet
STMicroelectronics
Crolles, France

Yong Han
Kairos Microsystems Corporation
Melrose, Florida

Guénolé Lallement
STMicroelectronics
Crolles, France

George Lentaris
National Technical University of
    Athens
School of ECE
Athens, Greece

Yan Lu
State Key Laboratory of Analog and
    Mixed-Signal VLSI
University of Macau
Macau, China

Konstantinos Maragos
National Technical University of
    Athens
School of ECE
Athens, Greece

Rui P. Martins
State Key Laboratory of Analog and
    Mixed-Signal VLSI
University of Macau

and

ECE Department
Faculty of Science and Technology
University of Macau
Macau, China

and

Instituto Superior Técnico
Universidade de Lisboa
Lisbon, Portugal

Nicole McFarlane
University of Tennessee
Knoxville, Tennessee

Philippe Roche
STMicroelectronics
Crolles, France

Emre Salman
Stony Brook University
Stony Brook, New York

Kostas Siozios
Aristotle University of Thessaloniki
Department of Physics
Thessaloniki, Athens

Dimitrios Soudris
National Technical University of
    Athens
School of ECE
Athens, Greece

Ioannis Stratakos
National Technical University of
  Athens
School of ECE
Athens, Greece

Liming Xiu
Kairos Microsystems Corporation
Melrose, Florida

Series Editor

Krzysztof (Kris) Iniewski is managing R&D at Redlen Technologies Inc., a startup company in Vancouver, Canada. Redlen's revolutionary production process for advanced semiconductor materials enables a new generation of more accurate, alldigital, radiation-based imaging solutions. Kris is also a founder of ET CMOS Inc. (http://www.etcmos.com), an organization of high-tech events covering communications, microsystems, optoelectronics, and sensors. In his career, Dr. Iniewski held numerous faculty and management positions at University of Toronto (Toronto, Canada), University of Alberta (Edmonton, Canada), Simon Fraser University (SFU, Burnaby, Canada), and PMC-Sierra Inc (Vancouver, Canada). He has published more than 100 research papers in international journals and conferences. He holds 18 international patents granted in the United States, Canada, France, Germany, and Japan. He is a frequently invited speaker and has consulted for multiple organizations internationally. He has written and edited several books for CRC Press (Taylor & Francis Group), Cambridge University Press, IEEE Press, Wiley, McGraw-Hill, Artech House, and Springer. His personal goal is to contribute to healthy living and sustainability through innovative engineering solutions. In his leisurely time, Kris can be found hiking, sailing, skiing, or biking in beautiful British Columbia. He can be reached at kris.iniewski@gmail.com.

About the Editor

Fei Yuan earned his BEng degree in electrical engineering from Shandong University, Jinan, China, in 1985; his MASc degree in chemical engineering; and his PhD degree in electrical engineering from the University of Waterloo, Canada, in 1999. During 1985–1989, he was a lecturer in the Department of Electrical Engineering, Changzhou Institute of Technology, Jiangsu, China. In 1989, he was a visiting professor at Humber College of Applied Arts and Technology, Toronto, Ontario, Canada, and Lambton College of Applied Arts and Technology, Sarnia, Ontario, Canada. He was with Paton Controls Limited, Sarnia, Ontario, Canada, as a controls engineer during 1989–1994. Since 1999, he has been with the Department of Electrical and Computer Engineering, Ryerson University, Toronto, Ontario, Canada, where he is a professor. Dr. Yuan served as the chair of the department during 2010–2015, and he currently serves as the director of quality assurance for the Faculty of Engineering and Architectural Science, Ryerson University. Dr. Yuan is the editor and a lead coauthor of *CMOS Time-Mode Circuits: Principles and Applications* (CRC, 2015); the author of *CMOS Circuits for Passive Wireless Microsystems* (Springer, 2010), *CMOS Active Inductors and Transformers: Principle, Implementation, and Applications* (Springer, 2008), and *CMOS Current-Mode Circuits for Data Communications* (Springer, 2007); and the lead coauthor of *Computer Methods for Analysis of Mixed-Mode Switching Circuits* (Kluwer, 2004). In addition, he is the author or coauthor of 10 book chapters and more than 220 refereed papers in international journals and conference proceedings. Dr. Yuan was awarded a Dean's Teaching Award (2016), a Ryerson Research Chair Award (2005), a Dean's Research Award (2004), early tenure (2003) from Ryerson University, a doctoral graduate scholarship from the Natural Science and Engineering Research Council of Canada (1997), a Teaching Excellence Award from Changzhou Institute of Technology (1988), and a Science and Technology Innovation Award from Changzhou municipal government (1988). Dr. Yuan is currently on the editorial board of a number of international journals, a fellow of the Institution of Engineering and Technology (IET), and a registered professional engineer in the province of Ontario, Canada.

1

Clock Generation and Distribution for Low-Power Digital Systems

Martin Cochet, Guénolé Lallement, Fady Abouzeid, and Philippe Roche

CONTENTS

1.1 Introduction

Digital circuits and systems have been a part of modern life for more than 40 years, whether in the form of digital signal processors, processors, or application-specific integrated circuits. All these systems rely on efficient flexible clock generation as well as distribution to pace their functionality and have some requirement in that process: the clock signal must be produced accurately (i.e., with a low jitter value), its frequency must be tunable to achieve different processing speeds and power trade-off, and the whole generation and distribution process must have a limited area and power overhead at the system level.

The last decade has seen a rise in low-power digital systems, from embedded mobile processors to Internet of Things (IoT) applications. When considering these systems, the previously described clock delivery constraints still exist partially, while some are alleviated and other problems arise.

The considered systems operate down to near-threshold voltages (0.4–0.6 V) and low frequencies (1–50 MHz) to fit within highly constrained power budgets (in the microwatt to milliwatt range). This translates to different clocking constraints: the absolute accuracy, often quantified as jitter in picosecond, is less critical over a longer clock period. Yet the power overhead of all parts of the clock chain must be compared to a much lower total power budget. This requires specific redesign strategies of the full clock chain. Last, low-voltage operation introduces increased variability and standard cell balancing issues that have to be accounted for.

This chapter describes the design strategies elaborated along the clock chain. The first section describes the typical topology of a low-power microcontroller system and introduces the clocking requirements. The second section introduces the reference clock sources. The third section presents frequency multiplier strategies, and the last section covers distribution through clock trees.

1.2 Different Levels of Timekeeping

This section focuses on the most power-constrained digital systems, for IoT applications. These systems assume different detailed architectures depending on their use cases, which range from miniaturized implantable biomedical sensors [1] to industrial machine diagnosis application [2], or the most commonly known personal IoT [3,4].

However different, these systems share a common structure, as illustrated in Figure 1.1. The system interacts with the physical world through sensors and actuators interfaced via an analog front end. Then, the data are processed by a microcontroller unit (MCU), or full processor, before being transmitted to a base station via a radiofrequency (RF) interface. The whole system is powered by a battery and/or the output of an energy harvester.

The different clock sources are shown in blue in Figure 1.1. First, an absolute time reference has to be generated, typically through a quartz on board, though on-chip integrated solutions are now proposed. This reference has a low frequency, with a standard of 32.768 kHz.* This reference has two roles.

---

* This frequency corresponds to 2^{15} Hz, making it easy to use for a 15-bit digital counter to generate 1 s time intervals.

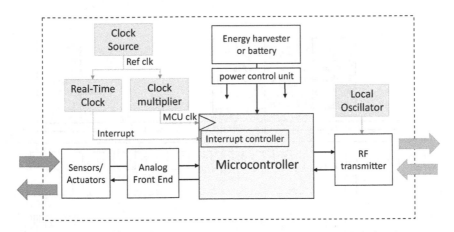

FIGURE 1.1
Generic block diagram of an IoT system.

First, it is used by the real-time clock (RTC) circuit, which acts as a counter to generate timed interrupt commands to the MCU. These are used to either acquire or transmit data at fixed intervals. Second, the same time reference can be used as an input to generate the main MCU clock. Depending on the application, that clock will operate from a couple to several hundred mega-hertz. A clock multiplier circuit is used to generate that signal. It is composed of an oscillator whose frequency is tuned to match a multiple of the reference. The most common architecture is a phase-locked loop (PLL), but other topologies will be presented in Section 1.4. Last, the RF transmitter uses a high-frequency local oscillator (e.g., 2.4 GHz for bluetooth). The specifications for this clock source are very different from the digital ones, and rely on different power management strategies, depending on the RF application, such as wake-up receivers [5], and are beyond the scope of this chapter.

Moreover, the IoT applications typically operate with a low amount of data to process. Hence, to fit within a constrained power budget, they operate on duty-cycled operation [6], as illustrated in Figure 1.2.

The digital clock plays a role at two different levels of timekeeping. The low-frequency RTC is used to dictate the phases of operation of the processor, in a time frame typically in the order of 1 s, while the multiplied clock generates the cycle-to-cycle operation of the processor. Hence, the performance requirements and specifications will be very different for the two. The RTC must have a very low drift over a long period of time to stay synchronous with the outside world or base station, while operating continuously, which requires a very low standby power. On the contrary, the MCU clock constraint is only to guarantee a low cycle-to-cycle jitter, but requires output flexibility, as it may need to be turned off or to operate at different frequencies (f1, f2 on Figure 1.2) depending on the MCU workload, a strategy named adaptive frequency scaling (AFS) [7].

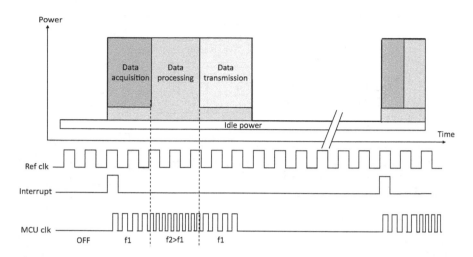

FIGURE 1.2
Duty-cycled operation of an IoT device, illustrating the different clocks used.

1.3 Time References

From the previous section, it is clear that low-power systems require precise time and frequency sources to ensure correct operations. However, while improving their precision, the designer should ensure a constant reduction in power, weight, and size of these sources to fit in the applications' power budgets. For the convenience of explanation, we will hereafter define a timer as any system that can be used as a clock source as well as an RTC or a reference for a clock multiplier.

After highlighting the design considerations related to time references in low-power digital systems, we will focus on some fundamental frequency and time metrology metrics. Lastly, a state-of-the-art analysis will be performed to provide insight into time reference trade-off according to the system constraints.

1.3.1 Design Considerations in IoT Nodes: Power, Area, and Stability

In order to bound the design considerations that must be explored during the conception phase, it is necessary to define some metrics and figures of merit (FoMs) for time references.

The power consumed by the timer to operate is the first characteristic that can be identified. In the specific case of a duty-cycled operation of an IoT device, designers gain room to shut off some IPs of the system-on-a-chip (SoC). However, the timer cannot benefit from that technique as it must always remain on for the entire lifetime utilization of the SoC. As a direct consequence, the regular power consumption of the timer may easily dominate the overall power budget compared with heavily duty-cycled high-power components such as the radio microprocessor [8].

With this in mind, the physical area of the clock should also be considered a key metric of any time reference. Indeed, the area footprint of a timer is directly related to the final production cost of the system. Smaller size enables higher volume of production of wafer-based timers. However, size improvement also impacts the device power dissipation. A smaller surface for heat dissipation into the environment leads to higher local variations that must be taken into account.

Moreover, as with any solid-state circuits, process, voltage, and environmental variations occurring during the fabrication or the operation of the timer will have an impact on its characteristics. Such variations often appear as a constant offset. Using negligible overhead logic, they can be easily corrected through calibration [9]. Likewise, in the case of low-power systems, circuits should operate at low voltages while tolerating voltage variations. A relative deviation of $\pm 10\%$ at 0.5 V results in a ± 50 mV change on the supply voltage that must also be taken into consideration.

Ultimately, various random sources of noise, such as thermal noise, will induce errors on the timer properties leading to a random spread of the frequency or period and phase. In the case where device synchronization is required between several timers, a random mismatch will appear. In that case, timing uncertainty will lead to added energy-expensive design margins.

1.3.2 Fundamental Frequency and Time Metrology

System frequency requirements are defined by the International Telecommunication Union (ITU) [10]. On the one hand, the accuracy measures the timer deviation value from a reference of the quantity being measured. On the other hand, the frequency change caused by any environmental and/or spontaneous action within a given time interval is expressed in terms of frequency stability. Stability can only be calculated on a set of measurement values as it describes a variation on measured samples. These accuracy and stability considerations are illustrated in Figure 1.3.

The time accuracy (i.e., period accuracy) and the frequency accuracy can be a single value or an average, and they are often normalized to a reference value. The dimensionless values $\frac{\Delta T}{T}$ and $\frac{\Delta f}{f}$ are generally used to describe time and frequency accuracy. In other words, accuracy expresses the proper setting of a timer on a target period or frequency reference. However, this metric does not indicate the variation around the reference set. Stability defines how well timers will keep the same accuracy over a given time

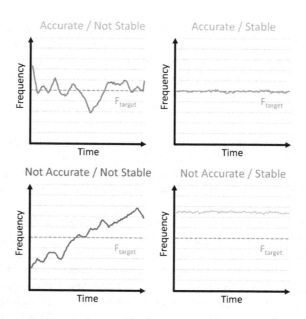

FIGURE 1.3
Accuracy and stability evaluation.

interval. Stability will not guarantee that the output period or frequency is accurate, but only that it remains the same. With this in mind, Figure 1.3 illustrates how an unstable device can be accurate while an inaccurate device can be temporarily stable.

In this section, we seek to define time references; thus, a common set of frequency stability characterization parameters is necessary. If a source guarantees a certain stability but presents inaccuracy, the last could always be seen as a constant offset and be removed. Therefore, in our application case, it is primordial to insist on the stability to ensure a proper time reference [11].

To ensure long-term stability, requirements on different applications can be described with two kind of parameters depending on the problem analysis domain selected. In the Fourier frequency domain, spectral parameters related to the spread of signal energy over the frequency spectrum are preferred. Spectral densities of phase and frequency fluctuation are particularly well adapted. In the time domain, the widely statistical time parameters used are variance or square root of the variance, also called standard deviation. Due to the inherent association between these two reciprocal domains, integral relationships exist to switch from spectral densities to variances, as shown in [12].

In the case of low-power time references, it is mandatory to characterize the stability of the timer over a time interval τ, which can range from milliseconds to years. In that circumstance, we prefer the time domain approach since it easily expresses the stability of the timer over a time interval of a given length. However, because of the standard variance's divergence due to the flicker noise, an M-sample variance was established by the early 1960s [13]. The use of the standard deviation stands on the existence of an absolute mean frequency, which is not a practical assumption, as shown in [14].

Therefore, as the IEEE standard *Definitions of Physical Quantities for Fundamental Frequency and Time Metrology* [15] recommends, we consider the timer's frequency stability through the utilization of the two-sample deviation $\sigma_y(\tau)$, also called Allan deviation. It is the square root of the two-sample variance $\sigma_y^2(\tau)$, also called Allan variance, and its expression is given in Equation (1.1) for a given observation time τ.

$$\sigma_y(\tau) = \left[\frac{1}{2} \left\langle [\bar{y}_{k+1} - \bar{y}_k]^2 \right\rangle \right]^{1/2} \tag{1.1}$$

The symbol $\langle \ \rangle$ denotes an infinite time average (i.e., $k \in [0; +\infty]$), whereas $\bar{y}_k$ is the kth instantaneous average of the fractional frequency deviation $y(t)$ over the time τ and defined by

$$\bar{y}_k = \frac{1}{\tau} \int_{t_k}^{t_k+\tau} y(t) \, \mathrm{d}t \tag{1.2}$$

where:
$t_k = t_0 + k\tau$ for some time origin t_0
$y(t)$ = normalized difference between the frequency $\nu(t)$ and the nominal frequency ν_n

The Allan variance expresses the variance of the timing error accumulated after a time interval τ relative to a reference clock, even if the mean oscillator's frequency is changing. This two-sample variance is a function of the sample period as it depends on the time period used between samples, contrary to the distribution being measured. Hence, as the sampling time must be reported, the Allan variance is generally displayed as a whole graph rather than a single value. A timer with good stability will exhibit a low Allan variance [16].

In practical applications, the infinite time average requirement cannot be fulfilled. Hence, we estimate the Allan deviation using the nonoverlapped estimate of the Allan deviation, where $y(t)$ being averaged over nonoverlapping intervals. For M number of frequency measurements, Equation (1.1) becomes

$$\sigma_y(\tau) \cong \left[\frac{1}{2(M-1)} \sum_{k=1}^{M-1} (\bar{y}_{k+1} - \bar{y}_k)^2 \right]^{1/2} \tag{1.3}$$

The application of such estimators in various cases is left to the curious reader and further explained in [17]. More derivations of the stability metrics in terms of frequency domains are given in this chapter's appendix. Based on these key metrics and stability metrology tools, we are now able to perform an overview of the available timing reference currently available.

1.3.3 Evaluation of Low-Power Time References

When it comes to generating an efficient time reference or clock source, plenty of techniques or circuits are available. Starting from common quartz oscillators to exotic chip-scale atomic time reference [18], designers currently have multiple options available depending on the design constraints of the application. Nonetheless, the different timers can be classified using their intrinsic principle of operation, as shown in Figure 1.4.

Linear oscillators, also known as harmonic oscillators, produce a sinusoidal output. In their most basic form, the output of a narrow-band electronic filter is amplified and then fed back into the input of the former filter while at the same time closing the loop of amplification. On the other hand, relaxation oscillators use a nonlinear component, like transistors or switching devices, to produce a nonsinusoidal output, such as saw-tooth or square waves.

There are numerous ways to implement both types of oscillator in order to create proper time references. However, recent ultra-low-power timers reported in the literature can be sorted into three types: crystal-based

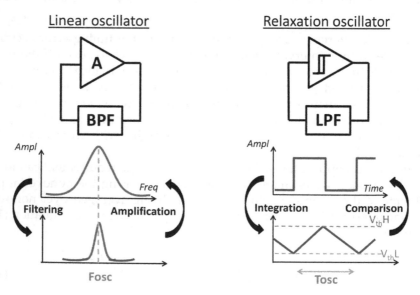

FIGURE 1.4
Oscillator classification based on their physical type of operation.

oscillators, complementary metal-oxide semiconductor (CMOS) circuits, and microelectromechanical system (MEMS) devices.

Crystal oscillators (XOs): These are widely known and used to generate reference clock signals onboard. The year 2018 marks the 100th anniversary of the first quartz oscillator [19]. They are based on the oscillation of a resonating piezoelectric element, such as silicon dioxide SiO_2, also called quartz. They are classified as linear oscillators with a narrow bandwidth due to the high-level Q-factor of the resonating element. Their oscillating frequency, which ranges from kilohertz to the hundreds of megahertz is directly related to the crystal physical dimensions. As a consequence, even though they can be easily calibrated one time through their dimensions, the mechanical behavior limits the run-time tuning range of a given unit.

With the proper calibration and control current, precision quartz oscillators, especially required in space applications, achieve an impressive stability of $\sigma_y(\tau) \in [5.10^{-14}, 7.10^{-14}]$ for τ from 1 to 100 s [20]. For common quartz crystals, lower values are expected ($\sigma_y(\tau) \approx 1.10^{-9}$), keeping them accurate and stable time references. However, despite the fact that such timers are very power efficient [21], reports a consumption of 1.5 nW for a 32.768 kHz crystal oscillator. Quartz timers are difficult to integrate because they require off-chip components. Moreover, their start-up time is quite long, from hundreds of microseconds in the megahertz range to a few seconds around 32 kHz [22].

CMOS-based circuits: In order to reduce the power, size, and SoC bill of materials (BoM) and to achieve full-scale integration, CMOS-based solutions have been developed within the last few years, as presented in [23]. Several techniques are available depending on the stability and degree of integration required.

Starting from the well-known ultra-low-power ring oscillators (ROs), a loop of an odd number of digital inverters, several solutions have been derived. Those timers that can be seen as a distributed version of the relaxation oscillator can easily be integrated using digital standard cells. However, due to their inherent structures, these types of oscillators tend to be highly sensitive to process, voltage, and temperature (PVT) variations, as shown in [24–26]. Especially when minimum-size digital logic transistors are used, they produce more output noise with limited supply noise rejection. To achieve even further low-power consumption, on-chip oscillators using gate leakage current have been proposed in [27,28]. However, the accuracy of such timers is not very well controlled across the fabrication process, leading to calibration requirements [29].

In order to overcome the inherent variability of the ring oscillator's circuit, relaxation circuits based on switching capacitors are also available. Without adding an impractical number of components and transistors, they often use compensation schemes or feedback loops, as in [30,31], to improve long-term stability and frequency stability over supply voltage, temperature, and process variations.

Lastly, today, linear CMOS solutions using LC resonators as the filtering element are ready for use. They currently achieve the most frequency stability of any on-chip timers. The passive LC components, as well as the transistors that provide the feedback amplification, can technically be integrated on chip. However, in order to produce low frequencies, a large LC couple is required. Passive elements can be used, but they will increase the size of the device. Otherwise, gyrator-based circuits that use active components reduce the area at the price of an increase in power consumption [32,33].

MEMS: Recently, new advances in micromachining techniques have led to development of MEMS sensors and timers. Depending on the technology node chosen, they can be integrated on chip, as demonstrated in [34]. Due to the different physical phenomena involved in their operation, MEMS resonators present a better accuracy and stability than CMOS circuits. Their power consumption and area are currently still higher than those of other timers, yet new circuit techniques and materials are offering perspectives for future applications. MEMS could be a trade-off between quartz and CMOS time references.

To offer an overview of the stability given by the aforementioned solutions, Figure 1.5 reports published work on time references and their Allan deviation with regard to their power consumption. The corresponding references associated with each point are given in Table 1.1.

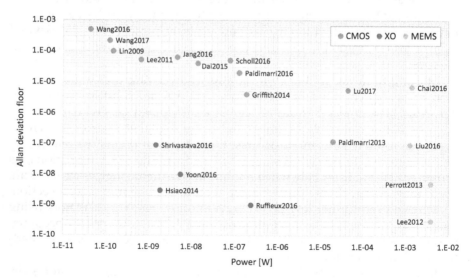

FIGURE 1.5
Allan deviation of timers with regard to their power consumption.

TABLE 1.1
Corresponding References and Data Associated with Figure 1.5

Label	Reference	Power [W]	Allan deviation floor $\sigma_y(\tau)$	Observation window τ [s]	Timer type
Scholl 2016	ESSCIRC'16 [28]	8.0×10^{-8}	5.0×10^{-5}	10	
Lin 2009	ISSCC'09 [35]	1.5×10^{-10}	9.9×10^{-5}	1400	
Lee 2011	ISSCC'11 [36]	6.6×10^{-10}	5.2×10^{-5}	1400	
Wang 2017	ESSCIRC'17 [37]	1.2×10^{-10}	2.2×10^{-4}	N/A	
Wang 2016	JSSCC'16 [38]	4.4×10^{-11}	5.0×10^{-4}	100	
Dai 2015	CICC'15 [39]	1.4×10^{-8}	4.0×10^{-5}	1	CMOS
Griffith 2014	ISSCC'14 [40]	1.9×10^{-7}	4.0×10^{-6}	10	
Lu 2017	VLSI'17 [41]	4.5×10^{-5}	5.5×10^{-6}	2	
Paidimarri 2013	ISSCC'13 [42]	2.0×10^{-5}	1.2×10^{-7}	100	
Jang 2016	JSSCC'16 [31]	4.7×10^{-9}	6.3×10^{-5}	100	
Paidimarri 2017	JSSCC'16 [43]	1.3×10^{-7}	2.0×10^{-5}	100	
Chai 2016	TENCON'16 [44]	1.4×10^{-3}	7.0×10^{-6}	1	
Lee 2012	IFCS'12 [45]	3.9×10^{-3}	3.0×10^{-10}	1	MEMS
Liu 2016	IFCS'16 [46]	1.3×10^{-3}	9.5×10^{-8}	1	
Perrott 2013	JSSCC'13 [47]	3.9×10^{-3}	5.0×10^{-9}	10	
Hsiao 2014	ISSCC'14 [48]	1.9×10^{-9}	3.0×10^{-9}	1000	
Shrivastava 2016	JSSCC'16 [21]	1.5×10^{-9}	9.0×10^{-8}	1000	XO
Yoon 2016	JSSCC'16 [49]	5.6×10^{-9}	1.0×10^{-8}	1000	
Ruffieux 2016	ISSCC'16 [9]	2.4×10^{-7}	1.0×10^{-9}	10	

1.4 Frequency Multipliers

After describing the original clock source, we need to consider how it can be scaled to a high frequency to clock the MCU, in a flexible and energy-efficient manner. This section introduces the quality metrics for such systems, reviews state-of-the-art low-power PLLs, and presents circuit architecture alternatives to PLLs.

1.4.1 Clock Multiplier Metrics

The clock multiplier circuit must be able to perform at low power and offer a fast response time to accommodate AFS strategies. This in turn requires the circuit to be integrated on chip. Indeed, an off-chip component would need too much area and system-level overhead (e.g., IO pin budgeting, extra board design) and would add latency to the frequency scaling. Moreover, as we will see, the jitter accuracy requirements are relaxed enough that a CMOS approach is viable, taking full advantage of advanced process scaling.

The *circuit area* must be controlled to limit the overhead compared with the full chip. A typical MCU would have an area in the order of square millimeters in a 28 nm node, so a reasonable area is in the order of 10,000–50,000 μm^2 (1%–5% overhead). To minimize skew and insertion delay, the clock generator should ideally be located on the core center or axis of symmetry. Hence, a smaller area makes floor planning easier. Last, a sub–$100 \times 100\ \mu m$ footprint can make possible the creation of finer-grained clock domains. This approach, named globally asynchronous, locally synchronous (GALS) network-on-a-chip (NoC) [50, 51], is a current research trend for energy-efficient SoCs.

The *frequency response* of the circuit corresponds to the transition time for the output to switch from a frequency f_1 to f_2. The target performance is constrained by the system-level specifications of the AFS algorithms. When the supply converter and controlled logic are fully integrated on chip, that time constant can be below 1 μs [52], that is, about 20 clock cycles at 50 MHz.

The *power consumption* and *jitter* are key metrics that are tied together when considered at the system level. The intrinsic power consumption of the clock generator can be defined relative to its generated frequency, in units of pJ/cycle (equivalent to 1 μW/MHz). The standby power consumption can also be considered, but is less critical than for the clock reference, as the clock multiplier only operates when the MCU is active and can be power gated in standby mode (see Figure 1.2). When considering clocking applications, the relevant jitter is the peak-to-peak cycle jitter, which corresponds to the maximum deviation of a given generated period compared with the target period, as illustrated in Figure 1.6. A more quantitative description of the cycle jitter can be found in this chapter's appendix. Typical circuits have a jitter of about 1%–5% UI, and a jitter $J = 2\%$ means that the period may be up to 1%

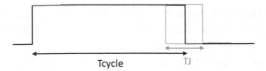

FIGURE 1.6
Definition of peak-to-peak period jitter; the relative jitter is defined as $J = Tcycle/T_J$.

early or 1% late. Hence, the MCU will need a 1% additional clock margin to account for clock uncertainty, which translates in a higher V_{DD} needed and higher power.

The *voltage and frequency range* of the clock generator are critical for voltage-scaled operations. First, in order to limit the number of supplies needed and to reduce SoC complexity, the clock multiplier should be able to operate on the same supply as the core logic. Moreover, a circuit operating at a fixed nominal voltage (e.g., 0.9 V) and corresponding maximum frequency (e.g., 1 GHz) would be able to produce the slower frequencies required by an MCU running at 0.5 V 50 MHz, but would result in poor efficiency in terms of energy per output cycle at 50 MHz.

1.4.2 Low-Power PLLs

The most popular circuits used for frequency multiplication are PLLs. These circuits operate by closed-loop control of an oscillator, so that its frequency-divided output stays in phase with an input reference [53]. PLLs are used ubiquitously for digital clock generation as well as RF, clock, and data recovery or deskewing applications.

The conventional analog PLL design uses a linear LC oscillator. Such circuits offer a very good jitter performance, which comes at the price of high area and limited frequency tunability because of the passive components. Designs using active LC components can partially mitigate those constraints at the price of higher power and jitter [54].

Hence, most compact low-power PLLs now use a digital implementation, replacing the analog building blocks with their digital counterparts. The phase detector is changed from a charge pump to a bang-bang or time-to-digital comparator. The loop filter can be synthesized from standard logic, and as the input data are only refreshed every reference clock cycle, the logic block can operate at a low frequency, making it simple to design and energy efficient. The oscillator is the most critical element of the digital PLL, as it dominates power and jitter performance. The most conservative approach is to use an analog LC oscillator with a digitally controlled varactor. Further area reduction and voltage scalability can be achieved by using a ring oscillator. A semianalog approach consists of implementing a voltage- or current-controlled oscillator (VCO/ICO) supplied by a digital-to-analog Converter

[55, 56]. A fully digital approach consists of using an array of tristate inverters on each stage of the oscillator: the frequency is tuned by controlling the fraction of the cells enabled [57, 58].

For low-power application, the first benefit of the digital substitutions is the supply scalability. Digital logic supply can be scaled down to subthreshold operation with only minor design efforts required [8, 59, 60], contrary to analog blocks. This offers a reduction in static and dynamic power consumption. The main trade-off of voltage scaling is the degradation of circuit speed. However, as those PLLs are designed to provide a clock signal to a digital circuit that is also scaled down in voltage and frequency, as long as the frequency degradation of the two circuits matches approximately, the PLL maximum frequency output will not impact system-level performance. The last benefit of such scaling at the system level is that the clock multiplier and the digital logic can potentially share the same supply, reducing the need for off-chip or on-chip regulators. Figure 1.7 summarizes published wide-voltage range clock multipliers with their maximum output frequency, including PLLs [56, 57, 61, 62] and alternative clock multipliers [6, 63–65], which will be introduced in the next section.

The second topic of research concerns fast-frequency-switching techniques. As PLLs are closed-loop circuits operating on a feedback control, a change in the input frequency or in the multiplication factor does not result in an immediate change in the output, but rather in a gradual relocking over several reference cycles. As explained previously, this can result in detrimental high latency for AFS schemes. In an even more extreme case, one wants

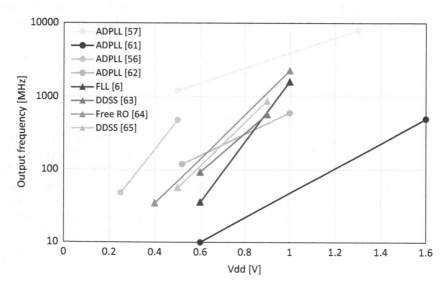

FIGURE 1.7
Comparison of supply range and frequency outputs of voltage-scalable clock multipliers.

to be able to fully power down a PLL (or at least turn off its main oscillator) and restart it to a known frequency in minimal time (see transition from *idle* to *data acquisition* stages in Figure 1.1).

The first approach used by PLLs to allow quick transitions from one frequency to another is to use direct frequency presetting [66–68]. When the target frequency is changed, the VCO/DCO control generated by the PFD is temporarily bypassed and set to the approximate expected value. During that step, the PLL temporarily performs in open loop. Then it resumes to normal operation to get to a fine-locked step, correcting, for example, for the oscillator nonlinearity. The most coarse approach consists of assuming a PVT independent linear model of the DCO and performing a one-step computation of the expected code to reach the target frequency. More refined methods have also been proposed to account for temperature [66] or DCO nonlinearities [67]. Another approach consists of cascading two circuits in a master-slave topology [68–70]. The master circuit is a conventional PLL, while the slave is either an injection-locked clocked multiplier or a fractional divider, offering fast open-loop tunability.

It is, however, worth noting that the quick phase-locking performance often has to be validated experimentally or with ad hoc simulations because of the complexity of the closed-loop response, especially when including the VCO's response speed [71,72].

1.4.3 Alternatives to PLLs

The main strengths of the PLL are the generation of a signal with a known phase relationship to a reference and a very low jitter. However, as described previously, those benefits are limited for low-power frequency generation, and they are outweighed by wide voltage range and fast-frequency-switching operation. Hence, different circuits have been proposed to offer a trade-off more suited to energy-efficient digital clocking.

The first option is to replace the PLL with a frequency-locked loop topology, (Figure 1.8(a)) [6,73,74]. This approach offers a significantly more compact area than a traditional PLL (1600 μm^2 in 28 nm process) and achieves a frequency response of about 60 ps. The reported jitter of 15 ps pk-pk for a maximum frequency of 2 GHz [74] is higher than conventional PLLs but still acceptable for clocking functions.

The second option is to use an open-loop oscillator for direct frequency generation (Figure 1.8(b)) [64,75]. This method is the simplest, offering a very low jitter, wide voltage range, low area, and instant frequency switching. It is also able to adapt instantly to voltage scaling without needing any active control: provided that the gates of the oscillator follow the delay of the logic critical path, any change in voltage, body biasing, temperature, or aging will be instantly transferred to the oscillator output period. However, this tracking is highly dependent on the clocked circuit and assumes a complete freedom of operating frequency. In the case where a specific

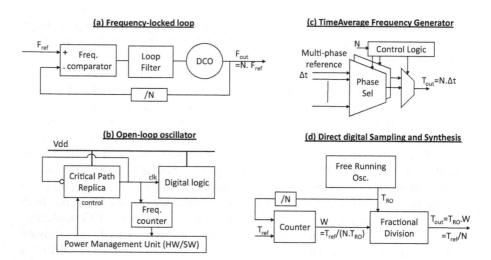

FIGURE 1.8
Alternative to PLL clock generator topologies.

frequency is constrained, a closed-loop (or open-loop lookup table) operation is necessary, resulting in an operation similar to that of a frequency-locked loop [52].

A third option, named time average frequency has been proposed [76,77] (Figure 1.8(c)). The general concept is that, in order to generate a frequency with a very fine granularity and fast switching, it is possible to alternate between two close frequencies, accepting a small jitter overhead. In practice, this circuit is implemented in a so-called flying-adder topology, where the two frequencies are obtained by phase selection of a main reference. The circuit-level implementation, however, requires a PLL providing the multiphase reference, which provides similar shortcomings to those of PLLs for voltage-scaled applications. Last, the benefit of this scheme in terms of extremely fine frequency resolution (in the order of parts per billion) is of limited use for digital logic clocking, contrary to communication applications, for example.

A last approach provides the open-loop nature of the ring oscillator while guaranteeing frequency locking to a reference. The topology is named direct digital sampling and synthesis (DDSS) (Figure 1.8(d)) [59,63,65]. The circuit includes a free-running ring oscillator acting as a time reference of period T_{RO}. The oscillator input, divided by a factor N, is first used to clock a simple counter, resetting every period of the slow reference T_{ref}. This way, the counter output is a digital word W equal to $W = N.T_{ref}/T_{RO}$. The output clock is produced by performing a fractional frequency division of the oscillator clock in a fashion similar to the time-average frequency generator. The main difference is that the division factor is not determined by a fixed input, but is equal to the output of the first-stage W. This way, the output period is equal to $T_{out} = T_{RO}.W = T_{RO}.N.T_{ref}/T_{RO} = T_{ref}/N$. The output frequency of the circuit is hence equal to exactly N times the reference frequency, independent of the exact value of the free-running oscillator period. This way, the oscillator does not have to be compensated for PVT variations. Moreover, because of the open-loop DDSS topology, the output value can be changed in one reference cycle when the multiplication value N is changed, or within a couple of output cycles if the input to the fractional divider W is overwritten by an external control.

The first version [59, 63] of the DDSS used delay lines to implement the fractional division stage, requiring extensive calibration to match the delay of the lines with that of the oscillator. The later proposed multiphase approach to the fractional division [65] resulted in a $14 \times$ area and $6 \times$ energy efficiency improvement. The DDSS is a novel and potentially interesting approach to low-power frequency generation. However, the published measurements report high absolute values of jitter, which may be detrimental for circuits operating at high frequency. Last, by construction, the oscillator period is only compensated every T_{ref}. So the accumulation of supply noise can lead to higher output jitter, compared with a regulated VCO, for example. Low-power circuits are less prone to IR drop or $L.di/dt$ noise, but tend to be noisier when including on-chip voltage converters such as buck boost [78] or switched capacitors [59, 79].

The published performance in terms of frequency and voltage range of current open-loop frequency multipliers is compared with that of PLLs in Figure 1.7.

1.5 Clock Distribution

Building an efficient clock network at low voltage requires specific solutions compared to nominal voltage, due to the new set of constraints induced by the gate delay and variability degradation [80, 81]. At nominal voltage, most of the effort is related to the network load and resistance. The resistance (R) and capacitance (C) of the clock network have a tremendous impact on a driver's capabilities in terms of clock signal propagation quality, that is, its ability to keep a balanced rise and fall delay to avoid any signal degradation. The electronic design automation (EDA) tools will be tuned to operate on the RC optimal metal layers, usually the highest ones, minimizing the length and buffering the high load by the distribution of buffers or inverters down to the leaves. The timing will be continuously checked to ensure that the time to reach all the leaves is close to equal, meaning reducing at best the skew between the longest and shortest paths in the network. Moreover, the tool will verify the clock pulse width when reaching the sequential cell, as it should remain wide enough for correct operation.

Lowering the voltage has two main impacts related to clock network operation: the strong gate delay and the variability increase, contrary to the RC effects, which do not scale with voltage, the delay induced by the wire being quite stable. The increased gate delay is coupled with a gentler transition time. The increased variability will therefore result in an extended range of possible transition times, producing an even wider range of gate delays. These behaviors make it even more complex to ensure the clock signal quality at the end of the network, or even predict the latency, therefore guaranteeing the clock network skew.

To reduce the variability window, a basic approach would be to increase the clock tree depth in terms of gate stages. But the increase of buffer cell count will drastically increase the power consumption. The buffering depth can be controlled efficiently through the maximum transition constraint. Basically, reducing the maximum transition value will induce a higher number of buffer stages as the transition time is related to the fan-out value. For a given value, the EDA software convergence will start to struggle, and the number of cells and gate stages will grow exponentially. The optimal constraint enabling steep transition, variability mitigation, and limited power consumption can be found at this point (Figure 1.9(a)). Another approach that benefits the variability mitigation is to limit the number of cell variants used for the clock tree design. Reducing the cell list to a single threshold voltage, gate length, and drive will mitigate the variability window compared with a wider set of possibilities, which would result in a less predictable window or averaging effect. Regarding the threshold voltage, using a lower threshold voltage compared with the logic will secure the sequencing as the variability on the clock tree skew will be lower (Figure 1.9(b)).

(a)   **(b)**

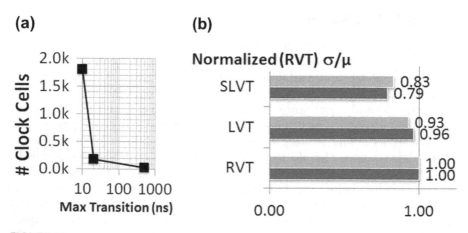

FIGURE 1.9
(a) Clock tree buffer count evolution with regard to clock transition constraint at low voltage.
(b) Delay variability of a ring oscillator at low voltage for different threshold voltages: super (S), low (L), and regular (R) threshold voltage (VT).

Last, designing a robust clock tree at low voltage can be achieved by using a single driver cell, which is able to control the load of the entire set of targeted leaf cells. The very low sensitivity in terms of delay of the RC network at low voltage offers, in this case, an extremely reduced skew. The drawback of this approach is that the operating range is limited to low voltage values, and the size of the design is constrained by the driver cell strength. The last point could be overcome using multiple drivers with their output shorted, but this would result in an increased power consumption due to the short-circuit state induced by the different response times of each cell.

Another effect of the voltage lowering is the discrepancy between the NMOS and PMOS current evolution across the voltage range. The result is the rise versus fall time unbalancing of the standard cells designed for nominal voltage operation. This could be improved using a specific set of cells designed for a given operating point.

Once designed, the clock tree may interact with other clock/voltage domains. The synchronism can be well secured relying on synchronization stages, correct constraint definitions, and the static timing analysis task. But a clock signal meant for being distributed in a given cycle should avoid crossing multiple voltage domains, as illustrated in Figure 1.10. First, each domain transition will impact the signal shape and timing. At best, the clock signal is generated and distributed in the same voltage domain. Otherwise, special care should be given to the cell in charge of the low-to-high or high-to-low voltage conversion in terms of rise versus fall balancing. Moreover, as insertion delay and variability are highly increased at low voltage, a crossed-branch tree (Figure 1.10, left) would have much harder-to-predict behavior than a dedicated-branch tree (Figure 1.10, middle) or a dedicated tree (Figure 1.10, right).

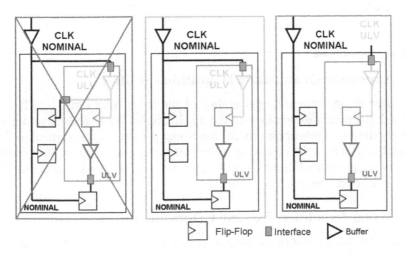

FIGURE 1.10
Clock tree constructions in the presence of multiple voltage domains.

1.6 Conclusion

Synchronous clocking operation has been the cornerstone of digital logic since the 1970s, enabling a high level of functional integration. Now that a power envelope constraint is part of virtually all systems, clocking has also played a role in dynamic power adaptation through AFS. However, the new generation of IoT application brings new challenges and constraints for SoC-level clock reference generation, flexible multiplication, and distribution.

An accurate time reference serves as the base of both on-chip operation and synchronization with other processing elements. Hence, the reference must offer an excellent long-term stability, as characterized by the Allan variance. Different references have been overviewed, with crystal oscillators offering the best accuracy, CMOS circuits improving integration, and MEMS sensors offering another booming field of devices.

The reference can also be used to produce the on-chip higher-frequency digital clock. Most often, the frequency multiplication is performed by a PLL. Recent research has improved their power and integration, as well as voltage scalability and fast frequency response. Several new substitutes for PLLs are also emerging, trading off jitter and phase/frequency locking for lower power.

Last, once the clock frequency is generated, it must be distributed to the MCU. At near-threshold voltages, the clock tree performance is dominated by logic gate devices rather than parasitic RC, and can be optimized by the use of a less deep clock tree at low voltage threshold.

IoT SoCs require an unprecedented level of integration and power reduction, which has researchers rethinking all the SoC building blocks. There is still a lot of innovation needed to provide low-power circuits all along the clock path.

A.1 Appendix: Quantitative Definitions of Jitter

The jitter characterizes the variability of the clock over time, which can be defined differently depending on the applications. We introduce and quantitatively define in this appendix the *long-term jitter*, which is critical to define the stability of time references, and the *period jitter*, which measures the error in a given generated clock period. Last, we discuss how the period jitter directly impacts the SoC power budget.

A.1.1 Long-Term Jitter: Frequency Domain Analysis

To understand the notion of long-term jitter, we introduce a simple model widely used for the instantaneous output voltage $v(t)$ of a precision oscillator [12–14]:

$$v(t) = (V_0 + \epsilon(t))sin(2\pi v_0 t + \phi(t)) \tag{A.1}$$

where:
V_0 is the nominal peak voltage amplitude
$\epsilon(t)$ is the deviation from the nominal amplitude
v_0 is the nominal frequency
ϕ_t is the phase deviation from the nominal phase $2\pi v_0 t$

Such a quasi-sinusoidal signal has a random frequency noise defined by

$$\Delta v(t) = v(t) - v_0 = \frac{1}{2\pi}\frac{d\phi(t)}{dt} \tag{A.2}$$

Generally, we introduce the instantaneous normalized frequency deviation $y(t)$ as follows to describe the frequency instability:

$$y(t) = \frac{\Delta v(t)}{v_0} = \frac{1}{2\pi v_0}\frac{d\phi(t)}{dt} \tag{A.3}$$

Therefore, the frequency stability of a timer can be expressed by the spectral density of the phase fluctuation $S_\phi(f)$ by

$$S_\phi(f) = \phi^2_{RMS}(f)\frac{1}{BW} \tag{A.4}$$

where:
$S_\phi(f)$ is in rad^2/Hz
$\phi_{RMS}(f)$ is the root mean square (RMS) value in the specific Fourier frequency bandwidth BW

Since phase and frequency are directly related, as shown in Equation (A.3), the spectral density of frequency can also be related to the phase fluctuation as follows:

$$S_y(f) = \frac{f^2}{v_0^2}S_\phi(f) \tag{A.5}$$

Two other quantities are also used to express the frequency stability: the jitter and the wander. According to the ITU, (timing) jitter and wander describe the long- or short-term variations from their ideal position in time of the significant physical parameter of a timing signal. Long-term refers to variation of frequency greater than 10 Hz, whereas short-term consists of variations below 10 Hz [10].

As shown in Figure A.1 and written in Equation (A.6), the RMS phase jitter deviation can be expressed by the square root of the integral of the spectral density of the phase deviation from a nominal phase between a low and a high cutoff frequency $f1$ and $f2$ [15].

$$\phi_{RMS}^{jitter} = \left[\int_{f_1}^{f_2} S_\phi(f)\, df \right]^{1/2} \tag{A.6}$$

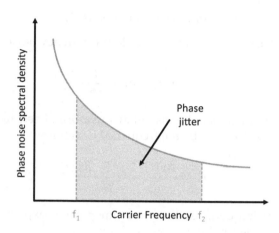

FIGURE A.1
Phase jitter and phase noise relationship.

Since the normalized or fractional frequency can be directly derived from the phase, specifying the phase noise is equivalent to specifying the frequency or period noise. Indeed, the angular frequency is the time derivative of the phase and shown in Equation (A.7)

$$\frac{d\phi(t)}{dt} = \frac{2\pi f(t)}{\nu_0} \tag{A.7}$$

As a consequence, at least for nearly sinusoidal signals, the RMS timing jitter J_{RMS} can be derived to the previous RMS phase noise as follows:

$$J_{RMS} = \frac{1}{2\pi\nu_0} \phi_{RMS}^{jitter} \tag{A.8}$$

From these derivations, we have seen that phase noise and jitter are interrelated and can both indicate the stability of a signal. However, it should be mentioned that the effect of frequency variation can also be performed in the time domain through peak-to-peak, cycle-to-cycle, and period jitter. In fact, these metrics compare the clock to itself to identify variations within the clock cycles. As they play a significant role in the clock multiplier metric, they are developed in the next section.

A.1.2 Period Jitter

Contrary to the long-term jitter, the period jitter only measures the deviation of a single period T_{out} compared with its ideal value, $N.T_{ref}$ in the case of a frequency multiplier.

Mathematically, the jitter J for a certain reference period T_{ref} and multiplication factor N is given by

$$J_{UI} = \frac{max(T_{out}) - min(T_{out})}{T_{ref}/N} \tag{A.9}$$

More specifically, the jitter will include a deterministic (e.g., device mismatch) bounded component and a Gaussian (e.g., device noise) unbounded component, expressed by its standard deviation, the RMS jitter. In practice, the jitter will be measured on a finite number of samples (e.g., 5000) giving a term equal to the sum of the deterministic and Gaussian components.

Last, RF applications often use the phase noise metric, which corresponds to the frequency domain distribution of the jitter (expressed in dBc/Hz). The period jitter can be derived from the integration of the phase noise spectral density, as described in [26, 82].

The jitter value directly impacts the timing margin T_{setup} needed to clock the MCU. Figure A.2 illustrates the effect of different values of jitter on the MCU period. The voltage and margins must be set in order to accommodate the minimum possible period T_{setup}, while the throughput of the MCU is

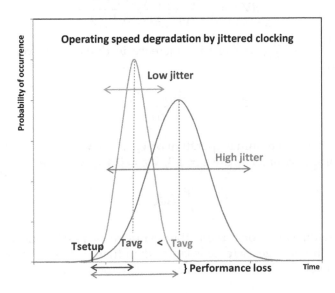

FIGURE A.2
Effect of jitter on an MCU operating frequency.

given by $T_{avg} = T_{setup}.(1 + J/2)$; that is, the throughput is degraded by a factor proportional to the jitter.*

A.1.3 Quantifying the Impact of Jitter on Power Budget

This throughput degradation can also be translated into a power overhead, corresponding to the extra voltage needed to guarantee operation at T_{setup} compared with the ideal T_{avg}. We propose an estimate of the impact of a degraded jitter on the full-circuit power budget, including analytical derivation in the case of super-threshold circuits and experimental data for near-threshold supply.

A simple estimate of CMOS logic speed with voltage can be given by the alpha power law [83]:

$$Fmax \propto \frac{(V_{DD} - V_{th})^\alpha}{V_{DD}}$$

So for a small jitter J, the relative frequency margin needed is $dF/F = J/2$ (a $J = 1\%$ jitter value corresponds to a $\pm 0.5\%$ error, and hence a 0.5% relative frequency margin is needed). Thus, we can derive

$$\frac{dF}{Fmax} = \left(\frac{\alpha}{V_{DD} - V_{th}} - \frac{1}{V_{DD}}\right) dV$$

$$dV = \frac{J/2.V_{DD}.(V_{DD} - V_{th})}{(\alpha - 1)V_{DD} + V_{th}}$$

As $P_{dyn} = CV_{DD}^2$, the relative increase in dynamic power is $dP/P = 2dV/V$, so

$$\frac{dP}{P} = \frac{(V_{DD} - V_{th})}{(\alpha - 1)V_{DD} + V_{th}} \times J$$

If we consider a nominal operation of $V_{DD} = 1$ V, $F_{max} = 1$ GHz, and that the technology parameters are $\alpha = 1.3$, $V_{th} = 0.4$ V, we have

$$\frac{dP}{P} \approx 0.86J$$

At near-threshold operation, this estimate is pessimistic because the speed-up with voltage increase is larger than that given by the alpha power law, so a lower dV/V is needed to compensate for the same dF/F. Moreover,

---

* This is assuming that the jitter distribution is symmetric. In the general case, the degradation is equal to $avg(T_{out}) - min(T_{out})$. An intermittent significantly larger period does not affect the functionality, whereas a single shorter period is enough to constrain T_{setup}.

the leakage component of the power can no longer be neglected. The relation can be determined experimentally by plotting power and frequency as a function of voltage; then for a given (F, V) point one can compute dF/dV and dP/dV, or directly dP/dF. Then we get

$$\frac{dP}{P} = \frac{dP}{dV} \cdot \frac{dV}{dF} \cdot \frac{dF}{F} \cdot F/P$$

$$\frac{dP}{P} = \left[\frac{1}{2} \frac{dP}{dV} \cdot \left(\frac{dF}{dV} \right)^{-1} \cdot F/P \right] \cdot J$$

Table A.1 compiles some experimental values found in the literature and the corresponding jitter/power relationship.

TABLE A.1
Experimental Data of Jitter and Power Overhead Relation

Reference	Vdd[V]	Power	Frequency (MHz)	(dP/P)/J
Analytical model [83]	1.0	—	1000	0.86
M0+ CPU [84]	0.5	15 μW	16	0.38
RISC-V CPU [52]	0.55	~45 GFLOPS/W	90	0.66
DSP [7]	0.5	33.4 mW	461	0.19

The experimental values are consistent between each other, showing that every increase in jitter by 1% UI corresponds to a 0.19%–0.86% increase in processor power. Last, the trade-off for the clock multiplier in terms of power versus jitter will depend on the absolute power of the full system compared with that of the clock generator. For example, in the case of [84], the overall absolute increase in power dP is only of 58 nW per 1% UI of jitter, that is, about 0.003 pJ/cycle. Hence, it is beneficial for the clock generator to trade more energy efficiency at the price of higher jitter. On the contrary, for more power-hungry systems such as [7], the power penalty is 137 μW, that is, 0.30 pJ/cycle for each increase of the jitter by 1%, making low jitter generation a higher priority.

References

1. X. Liu, M. Zhang, T. Xiong, A. G. Richardson, T. H. Lucas, P. S. Chin, R. Etienne-Cummings, T. D. Tran, and J. V. der Spiegel, "A fully integrated wireless compressed sensing neural signal acquisition system for chronic recording and brain machine interface," *IEEE Transactions on Biomedical Circuits and Systems*, vol. 10, no. 4, pp. 874–883, Aug. 2016.
2. "How to acquire machine diagnostics through energy harvesting in the industrial internet of things," Jun 2016. [Online]. Available: https://www.digikey.com/

en/articles/techzone/2016/jun/how-to-acquire-machine-diagnostics-through-energy-harvesting-in-the-industrial-internet-of-things.

3. J. A. Stankovic, "Research directions for the internet of things," *IEEE Internet of Things Journal*, vol. 1, no. 1, pp. 3–9, Feb. 2014.

4. C. Perera, C. H. Liu, S. Jayawardena, and M. Chen, "A survey on internet of things from industrial market perspective," *IEEE Access*, vol. 2, pp. 1660–1679, 2014.

5. C. Salazar, A. Kaiser, A. Cathelin, and J. Rabaey, "A -97dbm-sensitivity interferer-resilient 2.4 ghz wake-up receiver using dual-if multi-n-path architecture in 65 nm cmos," *2015 IEEE International Solid-State Circuits Conference (ISSCC) Digest of Technical Papers*, Feb. 2015, pp. 1–3.

6. D. E. Bellasi and L. Benini, "Smart energy-efficient clock synthesizer for duty-cycled sensor socs in 65 nm/28nm cmos," *IEEE Transactions on Circuits and Systems I: Regular Papers*, vol. 64, no. 9, pp. 2322–2333, Sept. 2017.

7. E. Beigne, F. Clermidy, D. Lattard, I. Miro-Panades, Y. Thonnart, and P. Vivet, "Fine-grain dvfs and avfs techniques for complex soc design: An overview of architectural solutions through technology nodes," *2015 IEEE International Symposium on Circuits and Systems (ISCAS)*, May 2015, pp. 1550–1553.

8. M. Keating, D. Flynn, R. Aitken, A. Gibbons, and K. Shi, *Low Power Methodology Manual: For System-on-Chip Design*, Springer, Berlin, 2007.

9. "DTCXO RTC module with an overall accuracy of compensation-resolution scheme at 1Hz," pp. 208–210, 2016.

10. International Telecommunication Union, "Definitions and terminology for synchronization networks," 1996.

11. "Ieee-Nasa Symposium on Short-Term Frequency Stability," *Nasa*, 1964.

12. L. S. Cutler and C. L. Searle, "Some aspects of the theory and measurement of frequency fluctuations in frequency standards," *Proceedings of the IEEE*, vol. 54, no. 2, pp. 136–154, 1966.

13. D. W. Allan, "Statistics of atomic frequency standards," *Proceedings of the IEEE*, vol. 54, no. 2, pp. 221–230, 1966.

14. J. A. Barnes, A. R. Chi, L. S. Cutler, D. J. Healey, D. B. Leeson, E. T. McGunigal, J. A. Mullen, W. L. Smith, R. L. Sydnor, R. F. C. Vessot, and G. M. R. Winkler, "Characterization of frequency stability," *IEEE Transactions on Instrumentation and Measurement*, vol. IM-20, no. 2, pp. 105–120, 1971.

15. IEEE Standards Coordinating Committee, *IEEE Std 1139-2008 (Revision of IEEE Std 1139-1999) IEEE Standard Definitions of Physical Quantities for Fundamental Frequency and Time Metrology–Random Instabilities*, 2009, vol. 2008, Feb.

16. D. W. Allan, "Time and frequency (time-domain) characterization, estimation, and prediction of precision clocks and oscillators," *IEEE Transactions on Ultrasonics, Ferroelectrics, and Frequency Control*, vol. 34, no. 6, pp. 647–654, 1987.

17. W. J. Riley, *Handbook of Frequency Stability Analysis*, 1994, vol. 31, no. 1. [Online]. Available: http://linkinghub.elsevier.com/retrieve/pii/0148906294927065.

18. R. Duggirala, A. Lal, and S. Radhakrishnan, "Radioisotope thin-film powered microsystems," *MEMS Reference Shelf*, vol. 6, p. 198, 2010.

19. D. W. Allan, N. Ashby, and C. C. Hodge, "The science of timekeeping," *Hewlett Packard Application Note 1289*, pp. 1–88, 1997. [Online]. Available: http://www.allanstime.com/Publications/DWA/Science_Timekeeping/TheScienceOfTimekeeping.pdf%5Cnhttp://literature.agilent.com/litweb/pdf/5965-7984E.pdf.

20. V. Candelier, P. Canzian, J. Lamboley, M. Brunet, and G. Santarelli, "Space qualified 5MHz ultra stable oscillators," *IFCS*, pp. 575–582, 2003. [Online]. Available: http://ieeexplore.ieee.org/lpdocs/epic03/wrapper.htm?arnumber=1275155.
21. F. V. Supply, A. Shrivastava, D. A. Kamakshi, S. Member, B. H. Calhoun, and S. Member, "A 1.5 nW, 32.768 kHz XTAL oscillator operational from a 0.3V supply," vol. 51, no. 3, pp. 686–696, 2016.
22. STMicroelectronics, "Application note oscillator design guide for STM8S, STM8A," pp. 1–24, 2015.
23. Y. Lee, Y. Kim, D. Yoon, D. Blaauw, and D. Sylvester, "Circuit and system design guidelines for ultra-low power sensor nodes," *Design Automation Conference (DAC), 2012 49th ACM/EDAC/IEEE*, pp. 1037–1042, 2012.
24. F. Herzel and B. Razavi, "A study of oscillator jitter due to supply and substrate noise," *IEEE Transactions on Circuits and Systems II: Analog and Digital Signal Processing*, vol. 46, no. 1, pp. 56–62, 1999.
25. V. Kratyuk, I. Vytyaz, U. K. Moon, and K. Mayaram, "Analysis of supply and ground noise sensitivity in ring and LC oscillators," *Proceedings — IEEE International Symposium on Circuits and Systems*, vol. 1, pp. 5986–5989, 2005.
26. A. A. Abidi, "Phase noise and jitter in cmos ring oscillators," *IEEE Journal of Solid-State Circuits*, vol. 41, no. 8, pp. 1803–1816, Aug. 2006.
27. A. Shrivastava and B. H. Calhoun, "A 150nW, 5ppm/°C, 100kHz on-chip clock source for ultra low power SoCs," *Proceedings of the Custom Integrated Circuits Conference*, pp. 12–15, 2012.
28. M. Scholl, Y. Zhang, R. Wunderlich, and S. Heinen, "A 80 nW, 32 kHz charge-pump based ultra low power oscillator with temperature compensation," *Proceedings of the European Solid-State Circuits Conference*, pp. 343–346, 2016.
29. D. Kamakshi, A. Shrivastava, C. Duan, and B. Calhoun, "A 36 nW, 7 ppm/°C on-chip clock source platform for near-human-body temperature applications," *Journal of Low Power Electronics and Applications*, vol. 6, no. 2, p. 7, 2016. [Online]. Available: http://www.mdpi.com/2079-9268/6/2/7.
30. S. Jeong, I. Lee, D. Blaauw, and D. Sylvester, "A 5.8nW, 45ppm/°C on-chip CMOS wake-up timer using a constant charge subtraction scheme," *Proceedings of the IEEE 2014 Custom Integrated Circuits Conference, CICC 2014*, pp. 3–6, 2014.
31. T. Jang, M. Choi, S. Jeong, S. Bang, D. Sylvester, and D. Blaauw, "A 4.7nW 13.8ppm/°C self-biased wakeup timer using a switched-resistor scheme," *Digest of Technical Papers — IEEE International Solid-State Circuits Conference*, vol. 59, pp. 102–103, 2016.
32. A. I. Karsilayan and R. Schaumann, "A high-frequency high-Q CMOS active inductor with DC bias control," *Midwest Symposium on Circuits and Systems*, vol. 1, pp. 486–489, 2000.
33. D. DiClemente and F. Yuan, "A passive transformer voltage-controlled oscillator with active inductor frequency tuning for ultra wideband applications," *Microsystems and Nanoelectronics*, no. 3, pp. 80–83, 2009. [Online]. Available: http://ieeexplore.ieee.org/xpls/abs{_}all.jsp?arnumber=5338956.
34. C. S. Lam, "A review of the recent development of mems and crystal oscillators and their impacts on the frequency control products industry," *Proceedings — IEEE Ultrasonics Symposium*, pp. 694–704, 2008.
35. Y. S. Lin, D. M. Sylvester, and D. T. Blaauw, "A 150pW program-and-hold timer for ultra-low-power sensor platforms," *Digest of Technical Papers — IEEE International Solid-State Circuits Conference*, pp. 326–328, 2009.

36. Y. Lee, B. Giridhar, Z. Foo, D. Sylvester, and D. Blaauw, "A 660pW multi-stage temperature-compensated timer for ultra-low-power wireless sensor node synchronization," *Digest of Technical Papers — IEEE International Solid-State Circuits Conference*, pp. 46–47, 2011.

37. H. Wang and P. P. Mercier, "A 1.6%/V 124.2 pW 9.3 Hz relaxation oscillator featuring a 49.7 pw voltage and current reference generator," *Proceedings of the European Solid-State Circuits Conference*, vol. 1, pp. 6–9, 2017.

38. H. Wang and P. P. Mercier, "A 51 pW reference-free capacitive-discharging oscillator architecture operating at 2.8 Hz," *Proceedings of the Custom Integrated Circuits Conference*, vol. 2015-Nov., pp. 0–3, 2015.

39. S. Dai and J. K. Rosenstein, "A 14.4nW 122KHz dual-phase current-mode relaxation oscillator for near-zero-power sensors," *Proceedings of the Custom Integrated Circuits Conference*, vol. 2015-Nov, pp. 4–7, 2015.

40. D. Griffith, P. T. Røine, J. Murdock, and R. Smith, "A 190nW 33kHz RC oscillator with 0.21% temperature stability and 4ppm long-term stability," *Digest of Technical Papers — IEEE International Solid-State Circuits Conference*, vol. 57, pp. 300–301, 2014.

41. S. Y. Lu and Y. T. Liao, "A 45µW, 9.5MHz current-reused RC oscillator using a swing-boosting technique," *2017 International Symposium on VLSI Design, Automation and Test, VLSI-DAT 2017*, 2017, pp. 5–8.

42. A. Paidimarri, D. Griffith, A. Wang, A. P. Chandrakasan, and G. Burra, "A 120nW 18.5kHz RC oscillator with comparator offset cancellation for ±0.25% temperature stability," *Digest of Technical Papers — IEEE International Solid-State Circuits Conference*, vol. 56, pp. 184–185, 2013.

43. A. Paidimarri, D. Griffith, A. Wang, S. Member, G. Burra, A. P. Chandrakasan, and A. A. Rc, "An RC oscillator with comparator offset cancellation," *IEEE J. Solid-State Circuits*, vol. 51, no. 8, pp. 1–12, 2016.

44. K. T. Chai, C. Wang, J. Tao, J. Xu, L. Zhong, and R. S. Tan, "High-performance differential capacitive mems sensor readout with relaxation oscillator front-end and phase locked loop time-to-digital converter back-end," *IEEE Region 10 Annual International Conference, Proceedings/TENCON*, pp. 1528–1531, 2016.

45. H. Lee, A. Partridge, and F. Assaderaghi, "Low jitter and temperature stable MEMS oscillators," *2012 IEEE International Frequency Control Symposium, IFCS 2012, Proceedings*, pp. 266–270, 2012.

46. C. Y. Liu, M. H. Li, C. Y. Chen, and S. S. Li, "An ovenized CMOS-MEMS oscillator with isothermal resonator and sub-mW heating power," *2016 IEEE International Frequency Control Symposium, IFCS 2016 — Proceedings*, pp. 4–6, 2016.

47. M. H. Perrott, J. C. Salvia, F. S. Lee, A. Partridge, S. Mukherjee, C. Arft, K. Jintae, N. Arumugam, P. Gupta, S. Tabatabaei, S. Pamarti, L. Haechang, and F. Assaderaghi, "A temperature-to-digital converter for a MEMS-based programmable oscillator with <±0.5-ppm frequency stability and <1-ps integrated jitter," *IEEE Journal of Solid-State Circuits*, vol. 48, no. 1, pp. 276–291, 2013. [Online]. Available: http://ieeexplore.ieee.org/ielx5/4/6399535/06341095.pdf?tp=&arnumber=6341095&isnumber=6399535.

48. K. J. Hsiao, "A 1.89nW/0.15V self-charged XO for real-time clock generation," *Digest of Technical Papers — IEEE International Solid-State Circuits Conference*, vol. 57, pp. 298–299, 2014.

49. D. Yoon, T. Jang, D. Sylvester, and D. Blaauw, "A 5.58 nW crystal oscillator using pulsed driver for real-time clocks," *IEEE Journal of Solid-State Circuits*, vol. 51, no. 2, pp. 509–522, 2016.

50. E. Beigne, F. Clermidy, H. Lhermet, S. Miermont, Y. Thonnart, X.-T. Tran, A. Valentian, D. Varreau, P. Vivet, X. Popon, and H. Lebreton, "An asynchronous power aware and adaptive NoC based circuit," *IEEE Journal of Solid-State Circuits*, vol. 44, no. 4, pp. 1167–1177, Apr. 2009.

51. E. Fluhr, J. Friedrich, D. Dreps, V. Zyuban, G. Still, C. Gonzalez, A. Hall, D. Hogenmiller, F. Malgioglio, R. Nett, J. Paredes, J. Pille, D. Plass, R. Puri, P. Restle, D. Shan, K. Stawiasz, Z. Deniz, D. Wendel, and M. Ziegler, "5.1 POWER8TM: a 12-core server-class processor in 22nm SOI with 7.6Tb/s off-chip bandwidth," *Solid-State Circuits Conference Digest of Technical Papers (ISSCC), 2014 IEEE International*, Feb. 2014, pp. 96–97.

52. B. Keller, M. Cochet, B. Zimmer, J. Kwak, A. Puggelli, Y. Lee, M. Blagojevic, S. Bailey, P. F. Chiu, P. Dabbelt, C. Schmidt, E. Alon, K. Asanovic, and B. Nikolic, "A risc-v processor soc with integrated power management at submicrosecond timescales in 28 nm fd-soi," *IEEE Journal of Solid-State Circuits*, vol. 52, no. 7, pp. 1863–1875, July 2017.

53. D. Fischette, "Practical phase-locked loop design," ISSCC tutorial, 2004. [Online]. Available: http://www.delroy.com/PLL_dir/ISSCC2004/PLLTutorialISSCC2004.pdf.

54. N. C. Shirazi, E. Abiri, and R. Hamzehyan, "A 5.5 GHz voltage control oscillator (VCO) with a differential tunable active and passive inductor," *International Journal of Information and Electronics Engineering*, vol. 3, no. 5, pp. 493–497, 2013.

55. I.-C. Hwang, S.-H. Song, and S.-W. Kim, "A digitally controlled phase-locked loop with a digital phase-frequency detector for fast acquisition," *IEEE Journal of Solid-State Circuits*, vol. 36, no. 10, pp. 1574–1581, Oct. 2001.

56. Y. Ho, Y.-S. Yang, C. Chang, and C. Su, "A near-threshold 480 MHz 78uW all-digital PLL with a bootstrapped DCO," *IEEE Journal of Solid-State Circuits*, vol. 48, no. 11, pp. 2805–2814, 2013.

57. J. Tierno, A. Rylyakov, and D. Friedman, "A wide power supply range, wide tuning range, all static CMOS all digital PLL in 65 nm SOI," *IEEE Journal of Solid-State Circuits*, vol. 43, no. 1, pp. 42–51, 2008.

58. Y. W. Chen and H. C. Hong, "A fast-locking all-digital phase locked loop in 90nm CMOS for gigascale systems," *2014 IEEE International Symposium on Circuits and Systems (ISCAS)*, June 2014, pp. 1134–1137.

59. S. Clerc, M. Saligane, F. Abouzeid, M. Cochet, J. M. Daveau, C. Bottoni, D. Bol, J. De-Vos, D. Zamora, B. Coeffic, D. Soussan, D. Croain, M. Naceur, P. Schamberger, P. Roche, and D. Sylvester, "8.4 a 0.33v/-40c process/temperature closed-loop compensation soc embedding all-digital clock multiplier and dc-dc converter exploiting fdsoi 28nm back-gate biasing," *2015 IEEE International Solid-State Circuits Conference — (ISSCC) Digest of Technical Papers*, Feb. 2015, pp. 1–3.

60. J. Myers, A. Savanth, R. Gaddh, D. Howard, P. Prabhat, and D. Flynn, "A sub-threshold ARM cortex-M0+ subsystem in 65 nm CMOS for WSN applications with 14 power domains, 10T SRAM, and integrated voltage regulator," *IEEE Journal of Solid-State Circuits*, pp. 31–44, 2016.

61. W. Liu, W. Li, P. Ren, C. Lin, S. Zhang, and Y. Wang, "A pvt tolerant 10 to 500 mhz all-digital phase-locked loop with coupled tdc and dco," *IEEE Journal of Solid-State Circuits*, vol. 45, no. 2, pp. 314–321, Feb. 2010.

62. C. C. Chung, W. S. Su, and C. K. Lo, "A 0.52/1 V fast lock-in ADPLL for supporting dynamic voltage and frequency scaling," *IEEE Transactions on Very Large Scale Integration (VLSI) Systems*, vol. 24, no. 1, pp. 408–412, Jan. 2016.

63. M. Cochet, S. Clerc, M. Naceur, P. Schamberger, D. Croain, J. L. Autran, and P. Roche, "A 28nm fd-soi standard cell 0.6-1.2v open-loop frequency multiplier for low power soc clocking," *2016 IEEE International Symposium on Circuits and Systems (ISCAS)*, May 2016, pp. 1206–1209.

64. J. Kwak and B. Nikolic, "A self-adjustable clock generator with wide dynamic range in 28 nm fdsoi," *IEEE Journal of Solid-State Circuits*, vol. 51, no. 10, pp. 2368–2379, Oct. 2016.

65. M. Cochet, S. Clerc, F. Abouzeid, P. Roche, and J.-L. Autran, "A 0.40 pj/cycle 981 um2 voltage scalable digital frequency generator for soc clocking," *2017 IEEE Asian Solid-State Circuits Conference (A-SSCC)*, Nov. 2017.

66. X. Kuang and N. Wu, "A fast-settling pll frequency synthesizer with direct frequency presetting," *2006 IEEE International Solid State Circuits Conference — Digest of Technical Papers*, Feb. 2006, pp. 741–750.

67. Y. W. Chen and H. C. Hong, "A fast-locking all-digital phase locked loop in 90nm cmos for gigascale systems," *2014 IEEE International Symposium on Circuits and Systems (ISCAS)*, June 2014, pp. 1134–1137.

68. S. Hong, S. Kim, S. Choi, H. Cho, J. Hong, Y. H. Seo, B. Kim, H. J. Park, and J. Y. Sim, "A 250-μW 2.4-ghz fast-lock fractional-n frequency generation for ultralow-power applications," *IEEE Transactions on Circuits and Systems II: Express Briefs*, vol. 64, no. 2, pp. 106–110, Feb. 2017.

69. A. Elkholy, A. Elshazly, S. Saxena, G. Shu, and P. K. Hanumolu, "A 20-to-1000MHz $\pm$14ps peak-to-peak jitter reconfigurable multi-output all-digital clock generator using open-loop fractional dividers in 65nm CMOS," *2014 IEEE International Solid-State Circuits Conference Digest of Technical Papers (ISSCC)*, Feb. 2014, pp. 272–273.

70. D. Coombs, A. Elkholy, R. K. Nandwana, A. Elmallah, and P. K. Hanumolu, "8.6 a 2.5-to-5.75ghz 5mw 0.3psrms-jitter cascaded ring-based digital injection-locked clock multiplier in 65nm cmos," *2017 IEEE International Solid-State Circuits Conference (ISSCC)*, Feb. 2017, pp. 152–153.

71. G. A. Leonov, N. V. Kuznetsov, M. V. Yuldashev, and R. V. Yuldashev, "Hold-in, pull-in, and lock-in ranges of pll circuits: Rigorous mathematical definitions and limitations of classical theory," *IEEE Transactions on Circuits and Systems I: Regular Papers*, vol. 62, no. 10, pp. 2454–2464, Oct. 2015.

72. L. Xiu, "Clock technology: the next frontier," *IEEE Circuits and Systems Magazine*, vol. 17, no. 2, pp. 27–46, 2017.

73. C. Albea, D. Puschini, P. Vivet, I. Miro-Panades, E. Beign, and S. Lesecq, "Architecture and robust control of a digital frequency-locked loop for fine-grain dynamic voltage and frequency scaling in globally asynchronous locally synchronous structures," *J. Low-Power Electronics*, vol. 7, pp. 328–340, 2011.

74. I. Miro-Panades, E. Beign, Y. Thonnart, L. Alacoque, P. Vivet, S. Lesecq, D. Puschini, A. Molnos, F. Thabet, B. Tain, K. B. Chehida, S. Engels, R. Wilson, and D. Fuin, "A fine-grain variation-aware dynamic *rmVdd*-hopping avfs architecture on a 32 nm gals mpsoc," *IEEE Journal of Solid-State Circuits*, vol. 49, no. 7, pp. 1475–1486, July 2014.

75. T. D. Burd, T. A. Pering, A. J. Stratakos, and R. W. Brodersen, "A dynamic voltage scaled microprocessor system," *IEEE Journal of Solid-State Circuits*, vol. 35, no. 11, pp. 1571–1580, Nov 2000.

76. L. Xiu, "The concept of time-average-frequency and mathematical analysis of flying-adder frequency synthesis architecture," *IEEE Circuits and Systems Magazine*, vol. 8, no. 3, pp. 27–51, 2008.

77. L. Xiu, *Nanometer Frequency Synthesis Beyond the Phase-Locked Loop*, IEEE Press Series on Microelectronic Systems, Wiley, Hoboken, NJ, 2012. [Online]. Available: https://books.google.com/books?id=e4hWNQAvkFUC.

78. W. Xu, Y. Li, Z. Hong, and D. Killat, "A 90dual-output buck-boost converter with extended-pwm control," *2011 IEEE International Solid-State Circuits Conference*, Feb. 2011, pp. 394–396.

79. Y. Ramadass, A. Fayed, B. Haroun, and A. Chandrakasan, "A 0.16mm2 completely on-chip switched-capacitor dc-dc converter using digital capacitance modulation for ldo replacement in 45nm cmos," *2010 IEEE International Solid-State Circuits Conference (ISSCC)*, Feb. 2010, pp. 208–209.

80. M. Seok, D. Blaauw, and D. Sylvester, "Clock network design for ultra-low power applications," *2010 ACM/IEEE International Symposium on Low-Power Electronics and Design (ISLPED)*, Aug. 2010, pp. 271–276.

81. J. R. Tolbert, X. Zhao, S. K. Lim, and S. Mukhopadhyay, "Analysis and design of energy and slew aware subthreshold clock systems," *IEEE Transactions on Computer-Aided Design of Integrated Circuits and Systems*, vol. 30, no. 9, pp. 1349–1358, Sept. 2011.

82. A. Demir, A. Mehrotra, and J. Roychowdhury, "Phase noise in oscillators: a unifying theory and numerical methods for characterization," *IEEE Transactions on Circuits and Systems I: Fundamental Theory and Applications*, vol. 47, no. 5, pp. 655–674, May 2000.

83. T. Sakurai and A. Newton, "Alpha-power law MOSFET model and its applications to CMOS inverter delay and other formulas," *IEEE Journal of Solid-State Circuits*, vol. 25, no. 2, pp. 584–594, Apr. 1990.

84. G. Lallement, F. Abouzeid, M. Cochet, J.-M. Daveau, P. Roche, and J.-L. Autran, "A 2.7pJ/cycle 16MHz SoC with 4.3nW power-off ARM Cortex-M0+ Core in 28nm FD-SOI," *2017 IEEE European Solid-State Circuits Conference (ESSCIRC)*, Sept. 2017.

2

Design of Low Standby Power Fully Integrated Voltage Regulators

Yan Lu and Rui P. Martins

CONTENTS

2.1 Introduction

Internet of Things (IoT) devices were projected to be installed in every object surrounding us. Most of the IoT sensing and communication circuits only operate for a very short amount of time and sleep or stand by for a much longer time, as shown in Figure 2.1. This very low-duty-cycle and high peak-to-average power ratio operation feature makes the ultralow standby power a critical specification for the IoT devices to prolong the battery life and recharging cycle.

Basically, there are three choices for the fully integrated voltage regulators: inductor-based DC-DC converter, switched-capacitor (SC) DC-DC converter, and linear regulator. Although the inductor-based converter

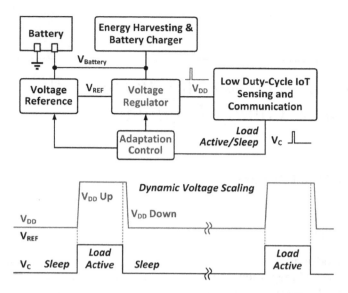

FIGURE 2.1
Block diagram of a low-duty-cycle IoT system with DVS.

can theoretically achieve 100% efficiency (while the other two cannot), it is not suitable for low-power applications. The required inductance of an inductor-based converter is inversely proportional to its load current. Therefore, when the load current is small, a large inductance is desired to maintain a small inductor current ripple and therefore a small peak-to-average current ratio. If the inductance is small, the converter would operate in discontinuous-conduction mode (DCM), which means the inductor current peak-to-average ratio would be high, increasing the I^2R conduction loss and reducing the power conversion efficiency. On the other hand, the required capacitance of an SC converter is linearly proportional to its output power, making the SC converter a favorable choice for low-power applications [1]. A linear regulator is a low-cost and easy solution. Because it does not have any energy storage component, the power transistor, which operates in an analog fashion, is the only element consuming chip area. Meanwhile, the controller power of the linear regulator can be designed as really small when compared with the load power [2]. Therefore, to address the low standby power issue of the fully integrated voltage regulators, we focus on the SC converter and the linear regulator designs in the following sections.

This chapter is organized as follows. Section 2.2 discusses the dynamic voltage scaling (DVS) power management scheme of the low-duty-cycle application. Section 2.3 introduces LDO regulator topologies and the low-power techniques with one design example. Section 2.4 addresses the SC DC-DC converters for low-power applications. Section 2.5 draws conclusions and proposes the future directions of this topic.

2.2 Power Management Scheme

As we know, the dynamic power of a digital circuit can be estimated by the following equation:

$$P_{DYN} = \alpha C V_{DD}^2 \cdot f_{CLK}, \tag{2.1}$$

where α is a switching activity factor, C is the total capacitance of the digital circuit, and f_{CLK} is the clock frequency. Thus, reducing V_{DD} has a square effect on the dynamic power. In addition, when V_{DD} is lower than $V_{TP} + V_{TN}$, where V_{TP} and V_{TN} are the threshold voltages of PMOS and NMOS transistors, respectively, the complementary metal-oxide semiconductor (CMOS) transistors in digital circuits will never be turned on simultaneously, and thus the short circuit current in the logic cells will be significantly reduced. Therefore, subthreshold operation with minimum energy consumption has been well studied in the past decade [3, 4]. From the energy per operation perspective, with the equation given below,

$$E_{OP} = P_{DYN} \cdot T_{OP} = \alpha C V_{DD}^2 \cdot f_{CLK} T_{OP}, \tag{2.2}$$

if f_{CLK} has a 10-fold reduction, the operation time T_{OP} would increase 10-fold as well. That means the consumed energy for the same computation task remains the same as $\beta C V_{DD}^2$, where $\beta = \alpha f_{CLK} T_{OP}$. While β and C are both constant, to reduce the energy consumption per operation, V_{DD} should be lowered to the minimum value that can satisfy computation speed requirements.

Therefore, to reduce the power consumption for different load demands or in the standby (or sleep) mode, and to maintain the minimum required performances in the active mode, DVS and adaptive voltage scaling (AVS) schemes were employed to manage the supply voltages and power consumptions [5, 6]. The difference between DVS and AVS is that DVS uses predefined V_{DD} values for targeted load demands and clock frequencies, while AVS is an advanced version that employs closed-loop V_{DD} control with real-time information of the load demand and on-chip temperature. Moreover, AVS would automatically calibrate the V_{DD} for the process variations among different chips. To realize AVS, dedicated hardware including replica load circuitry and temperature sensors should be integrated [7].

Usually, the DVS scheme is obtained by switching converters (inductor-based or SC DC-DC converter) that can efficiently step down the supply voltage [8–13]. With an ideal switching converter, $V_{IN} \cdot I_{IN} = V_{OUT} \cdot I_{OUT}$. Lowering V_{OUT} (V_{DD} of the digital load) will result in lower I_{OUT}; thus, the input power can be significantly reduced. However, the switching converters need bulky off-chip inductors or large on-chip capacitors (area) to temporarily store the energy during the power conversion. For some IoT devices that have stringent size and cost constraints, a low-dropout (LDO) regulator can also be an easy solution for DVS. Although the LDO regulator always draws $I_{IN} = I_{OUT} + I_Q \approx I_{OUT}$ no matter how low the V_{OUT} is, where I_Q is the quiescent current of the LDO regulator and is usually negligible compared with I_{OUT}, its input power ($V_{IN} \cdot I_{IN} \approx V_{IN} \cdot I_{OUT}$) can still be reduced with a lower V_{OUT} because, as mentioned above, a lower V_{OUT} reduces I_{OUT} for a digital load. Then the E_{OP} of a digital load with an LDO supply is $V_{IN} \cdot I_{OUT} \cdot T_{OP} = V_{IN} \cdot \beta C V_{DD}$, which decreases linearly with respect to V_{DD}.

Let's revisit Figure 2.1, which shows a voltage regulator with a voltage reference and an adaptation block controlled by the low-duty-cycle load. When the load intends to wake up and boost the performance, a digital signal V_C is sent to the adaptation block, which sets the reference voltage V_{REF}, bias current, and operation clock frequency of the voltage regulator. Then V_{REF} ramps up and the regulator output V_{OUT} tracks V_{REF}. Also, to provide a fast reference tracking capability, the bias current or the operation frequency of the voltage regulator should be boosted up together with V_{REF}. After the required tasks are finished in the active mode, the load would toggle the V_C and turn down V_{REF} and the biases immediately to save energy.

Now, a question remains whether V_{OUT} should be freely discharged from V_{O1} to V_{O2} by the load current I_O (case 1 in Figure 2.2), or V_{OUT} should be intentionally pulled down with additional current I_{PD} to reduce the dynamic power of the load in the t_1 to t_2 region (case 2 in Figure 2.2). Which case is more energy-efficient?

In the following analyses, we assume that the digital load can continuously function well with a slowly decreasing V_{DD}, and for simplicity, the I_O is constant. At time = 0, V_{REF} decreases from V_{O1} to V_{O2}. The energy consumptions E_1 and E_2 during the voltage step-down processes of both case 1 and case 2 are calculated as follows:

$$E_1 = \tfrac{1}{2}\left(V_{O1} + V_{O2}\right) I_O t_2, \tag{2.3}$$

$$E_2 = \tfrac{1}{2}\left(V_{O1} + V_{O2}\right)\left(I_O + I_{PD}\right) t_1 + V_{IN} I_O \left(t_2 - t_1\right). \tag{2.4}$$

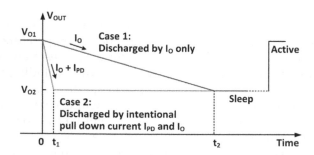

FIGURE 2.2
DVS schemes with case 1, V_{OUT} being discharged by I_{PD} and I_O, and with case 2, discharged by I_O only.

During t_1 to t_2 in case 2, V_{OUT} has been discharged to be equal to V_{REF}; thus, the power is supplied by V_{IN} in this period, as expressed in Equation (2.4). On the other hand, for case 1, the energy stored on C_L provides all the power consumed from 0 to t_2. Since both cases discharge C_L by the same ΔV, one more equation can be obtained:

$$C_L \Delta V = (I_O + I_{PD}) t_1 = I_O t_2. \tag{2.5}$$

Then, by substituting (2.5) into (2.4), we'll have

$$E_2 = \tfrac{1}{2} (V_{O1} + V_{O2}) I_O t_2 + V_{IN} I_O (t_2 - t_1), \tag{2.6}$$

$$E_2 - E_1 = V_{IN} I_O (t_2 - t_1). \tag{2.7}$$

Thus, case 2 consumes more energy than case 1 because it needs to draw energy from the input source during the t_1 to t_2 period. Therefore, case 1 should be adopted in the DVS scheme for ultra-low-power energy-limited IoT devices.

2.3 Low-Dropout Regulators

Figure 2.3 shows a typical LDO regulator with the parasitic N-Well/P-Sub junction capacitors and the CMOS gate capacitors in the load circuit, which also serve as the load capacitor C_L. For the near-threshold digital circuits, relatively larger transistor widths are used in the logic cells to compensate for the performance loss of subthreshold operation. In a system-on-a-chip (SoC), these parasitic capacitors that contributed to the C_L of the voltage regulator can be in the range of 1 pF to 10 nF. Also, when the reverse bias voltage of the junction diodes changes (e.g., from 1 V to 0.5 V), there we may have a certain variation of the capacitance. That is one of the reasons why the output-capacitor-free LDO regulators need to guarantee the stability over a wide load capacitance range.

We need capacitors for filtering and compensation, but at the same time, they limit the bandwidth of an analog circuit. For an LDO regulator, the largest capacitors are the load capacitor C_L and the parasitic gate capacitor C_g of the power MOS transistor. Hence, there are at least two low-frequency poles on the left-half-plane (LHP): the pole at the output node, $-p_{Out}$, and the pole at the gate of the power MOS, $-p_{Gate}$, as sketched in Figure 2.4(a) and (b), with either p_{Out} or p_{Gate} being the dominant pole. Also, there might be multiple low-frequency poles in the error amplifier (EA) of the LDO regulator, when two or more amplifier stages are employed for higher loop gain.

Conventionally, to have an ultra-low-power design, it is suggested to design the dominant pole at one of the internal nodes of the LDO regulator.

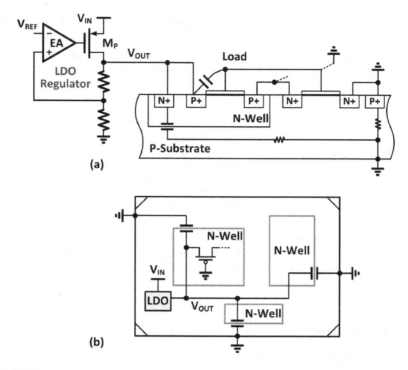

FIGURE 2.3
(a) Generic LDO regulator driving a digital load with parasitic load capacitors. (b) Top view of the system chip.

This is because reducing the quiescent current is favorable to lowering the frequencies of the internal poles, while designing p_{Out} as the dominant pole requires relatively large current to push the internal poles to high frequencies [14–16]. However, the internal-pole dominant case may fail at the large C_L and light load case, in which p_{Out} comes into the unity-gain frequency (UGF), degrading the stability. In the past decade, a number of compensation techniques have been proposed for the analog LDO regulators, with internal pole being their dominant pole to achieve a wide load current and capacitance range [17–21]. Meanwhile, to further reduce the quiescent current at light load conditions, adaptive biasing techniques can be used without sacrificing the dynamic performance at heavy load conditions [22–24].

Regardless of the drawback of consuming more current, there are several benefits in designing p_{Out} as the dominant pole by using most of the available capacitance (area) at the output node. First, a larger output capacitor filters out power supply noise and glitches and serves as a buffer for load-transient current changes, resulting in a smaller ΔV_{OUT}. Second, as Figure 2.4(c) and (d) shows, because the output voltage is well regulated by the control loop at low frequency, and the noise is bypassed to ground by C_L at high frequency, the worst-case power supply ripple rejection (PSRR) would occur at medium

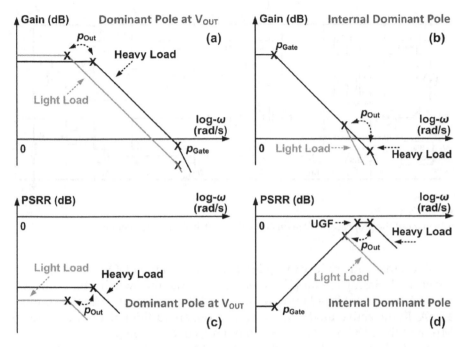

FIGURE 2.4
Sketched open-loop gain plots of an LDO regulator with dominant pole (a) at V_{OUT} or (b) at an internal node. PSRR of the case with dominant pole (c) at V_{OUT} or (d) at an internal node.

frequency [25]. Thus, increasing both the output capacitance and the loop bandwidth would improve the PSRR. Third, as the load current decreases, p_{Out} moves to lower frequencies, and it would be easier to maintain the loop stability compared with the internal-pole dominant case. Nevertheless, when adaptive biasing for the LDO regulator is enabled by internal or external configurations, ultralow power, fast transient response, and good PSRR performances can still be achieved simultaneously for the p_{Out} dominant case, as demonstrated by the following design example.

2.3.1 LDO Regulator with Switched Biasing

Based on the above analyses, an LDO regulator design example with switched biasing and adaptive output voltage is implemented in a low-leakage (LL) 65 nm CMOS process. Figure 2.5 shows the schematic of the LDO regulator with a switched-biasing technique. It consists of a differential input stage with current mirror loads (M_1–M_4), a g_m-boosting stage (M_5 and M_6) [16], a gain stage (M_7), a super source follower buffer stage (M_8–M_{10}), and a switched-biasing circuit (M_{11} and its left part) controlled by the low-duty-cycle load.

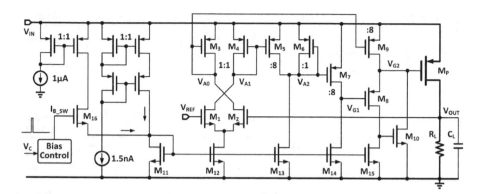

FIGURE 2.5
Schematic of the proposed LDO regulator with a switched-biasing technique.

To dramatically reduce the quiescent current of the LDO regulator, we use a switched-biasing circuit [2]. In the sleep mode, the switch M_{16} is off, only 1.5 nA bias current is fed into M_{11}, and the total quiescent current I_Q is about 45 nA. In the active mode, M_{16} is turned on, an additional 1 μA is added to M_{11}, and the LDO regulator consumes an I_Q of about 28 μA.

In this design, the timing of switching M_{16} on/off is generated by a timing control circuit, as shown in Figure 2.6. When the control signal V_C from the low-duty-cycle load comes to the LDO regulator, the turn-on edge will be immediately passed to the bias circuit with only two negligible logic delays. On the contrary, the turn-off edge will be delayed roughly 300 ns by the low-power delay and logic cell. This is to prolong the large bias period of the LDO regulator for a while, thus reducing the unwanted voltage overshoot during the switching off of the load. When codesigning the power management circuit with the load, this turn-off delay can be realized by one or a few clock cycles of the digital load instead.

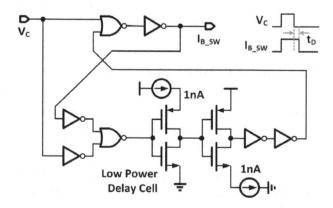

FIGURE 2.6
Schematic of the switched-biasing control circuit.

We design the dominant pole at the output node for better PSRR and higher control loop bandwidth, as suggested in [15]. Also, a large capacitive load (C_L = 1 nF in this design) from the near-threshold digital circuits that naturally sets the output pole to low frequency stabilizes the loop more easily. To lift the frequencies of the internal poles, we use a current mirror load in the differential input stage; on the other hand, we also utilize a super source follower that provides low input capacitance and output impedance to increase the frequencies of the poles on V_{G1} and V_{G2} simultaneously. This can be further improved as mentioned in [16]. The g_m-boosting stage provides additional gain without introducing any additional low-frequency pole to the loop.

The load current range of the LDO regulator for the active mode is from 0 nA to 2 mA. Figure 2.7 shows the frequency response with 28 μA I_Q. The worst-case (I_O = 2 mA) DC gain and phase margin are 39 dB and 48°, respectively. The maximum UGF is 90 MHz. Figure 2.8 shows the frequency response of the LDO regulator when it operates in the sleep mode with 45 nA I_Q. The worst-case DC gain happens at I_O = 200 nA, because V_{G1}

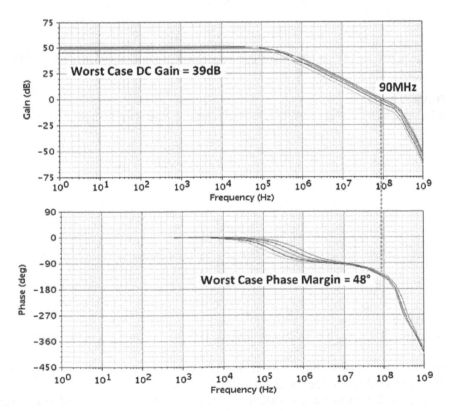

FIGURE 2.7
Simulated frequency response of the switched-biasing LDO regulator in the active mode.

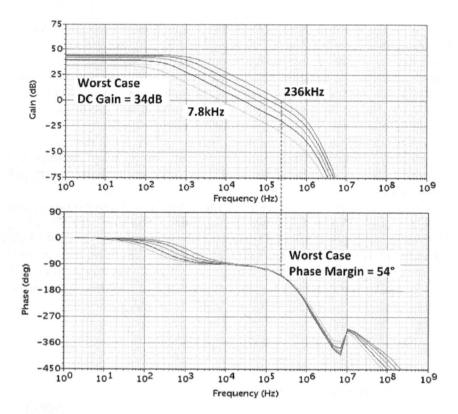

FIGURE 2.8
Simulated frequency response of the switched-biasing LDO regulator in the sleep mode.

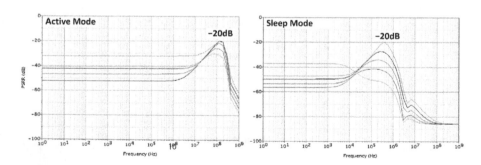

FIGURE 2.9
Simulated PSRR of the switched-biasing LDO regulator in either the active or the sleep modes.

and V_{G2} are approaching V_{IN} at this ultralight load and ultralow I_Q case. Then the load range for the sleep mode is from 200 nA to 2 μA.

Figure 2.9 shows the simulated PSRR of both the active and the sleep modes. Both cases can provide a −20 dB worst-case full-spectrum PSRR within the load ranges. Figure 2.10 gives the simulated transient performance

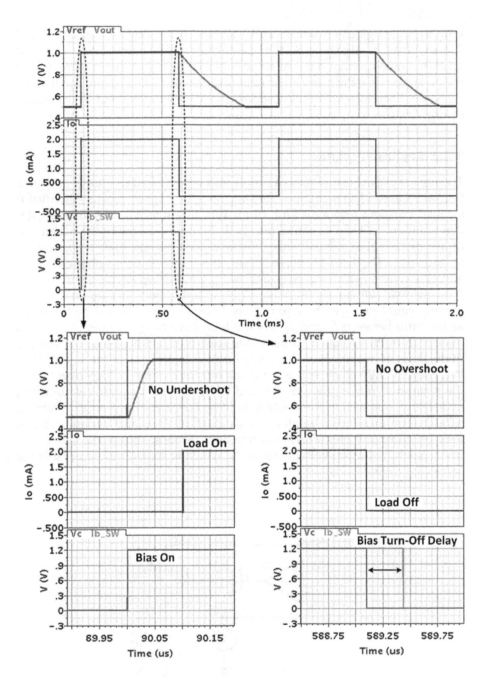

FIGURE 2.10
Simulated mode and load transitions of the switched-biasing LDO regulator.

with V_{REF} changing between 0.5 and 1 V, and I_O changing between 2 μA and 2 mA with a 1 ns edge time. As the bias current will be immediately switched on at the wake-up edge, a fast reference uptracking of 11 V/μs is achieved. At the go-sleep edge, the bias will be kept on for about 300 ns to maintain a fast response. Therefore, we don't observe any voltage undershoot or overshoot during the load transient.

2.3.2 Replica Regulator

The replica biasing technique is a simple and widely used technique in source follower–based LDO regulators, for supplying the digital circuits with intrinsic load-transient responses [26], or for realizing internal rails for the gate-drive buffers in DC-DC converters [13]. Figure 2.11 shows the schematic of a typical replica regulator that has a replica biasing branch formed by I_B and M_{N1}. The EA only senses V_{MIR} that is the source voltage of M_{N1} and generates V_G for both M_{N1} and M_{N2}, while M_{N2} supplies the current I_L to the load. Obviously, to make V_{OUT} exactly equal to V_{MIR} and consequently equal to V_{REF}, the size ratio of M_{N2} and M_{N1} should be the same as the ratio between I_L and I_B. As the load current I_L changes, V_{OUT} will deviate from the designed value. When load transient happens, the voltage undershoot or overshoot of V_{OUT} will automatically change the V_{GS} of M_{N2} and consequently changes the regulator output current. If M_{N2} operates in the saturation region, the voltage and current have a square relationship, and if M_{N2} operates in the subthreshold region, they have an exponential relationship; therefore, a small ΔV_{OUT} can result in a large output current. This structure provides a fast load-transient response, but at the same time it sacrifices the output voltage accuracy, which is sometimes acceptable for supplying digital loads.

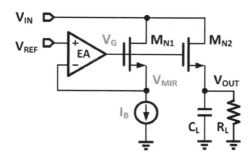

FIGURE 2.11
Schematic of the replica LDO regulator.

2.3.3 Flipped Voltage Follower–Based LDO Regulator

The flipped voltage follower (FVF) [27]–based LDO regulator is one of the most popular architectures due to its simplicity and the potential for a fast transient response [14, 19, 20]. Figure 2.12 shows the schematic of a single-transistor-controlled LDO regulator [19] based on the FVF topology as an example. This circuit can be divided into three parts: the EA, the V_{SET} generation, and the FVF. For simplicity, we assume $I_1 = I_2$ and $(W/L)_7 = (W/L)_8$. The mirrored voltage V_{MIR} is controlled by the EA to be equal to V_{REF}, and V_{SET} is generated from V_{MIR} by the diode-connected transistor M_7. Followed by an FVF stage, V_{OUT} is set by V_{SET} through M_8, and it is a mirrored voltage of V_{MIR}. In the FVF stage, M_8 also acts as a common-gate amplifying stage from V_{OUT} to V_G. Therefore, compared with the traditional source follower discussed in Section 2.3.2, the FVF has a higher voltage-to-current transconductance, at the expense of additional loop stability considerations, as discussed next.

Obviously, there are two low-frequency poles (p_G and p_{Out}) in the FVF loop when a relatively large load capacitor C_L needs to be driven. It is very difficult (if not impossible) for this topology to be stable if p_{Out} is the dominant pole. Another issue associated with this structure is the DC accuracy of V_{OUT}. The offset voltage between V_{REF} and V_{OUT} can be divided into two parts. First, there is an offset between V_{REF} and V_{MIR} that consists of systematic and random offsets of the EA. Second, the mismatches between the voltage mirror (M_7 and M_8) and the bias currents (I_1 and I_2) will generate an offset between V_{MIR} and V_{OUT}. Hence, the FVF-based topology has low immunity to the process, voltage, and temperature (PVT) variations. More-

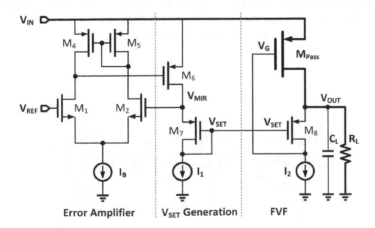

FIGURE 2.12
Schematic of an FVF-based LDO regulator.

over, the loop gain of the simple FVF is low, which results in poor load regulation, and tens of millivolts of V_{OUT} variations can be easily observed due to the load current change. To improve the DC accuracy, a folded-cascode FVF topology [16, 20] or a triloop topology [14, 15] providing higher loop gain can be used.

Since the FVF-based LDO regulator is a single-ended topology, for similar dynamic performance, the FVF-based LDO regulator only consumes half of the bias current compared with a conventional differential input LDO regulator [16]. Although the FVF-based LDO regulator also consists of an auxiliary EA with a differential input stage, it is not in the main loop, and only serves as a bias voltage generator, consuming only a negligible current. Thus, the FVF-based LDO regulator is a relatively power-efficient solution.

2.3.4 Digital LDO Regulator

In low-voltage scenarios, the voltage headroom for analog EA is very limited, and therefore the DC gain would be low. Alternatively, digital LDO regulators are popular for their easy reconfigurability and low-voltage features [28–33]. Figure 2.13 shows the schematic of a conventional digital LDO regulator. The digital LDO regulator employs one clocked comparator, which can be considered a 1-bit analog-to-digital converter (ADC); one bidirectional shift-register array; and one power transistor array. The $D[1:n]$ with unary code controls the number of unit transistors to be turned on and consequently the total output current. The comparator compares V_{REF} and V_{OUT} in every clock cycle to decide whether the shift-register output bits $D[1:n]$ shift to the left (add a 1) or to the right (add a 0). Obviously, the shift register acts as an integrator in the control loop and results in a pole at DC. The clocked comparator can operate at low supply voltages and only consumes power at the clock edges, while other digital cells can operate at low voltage as well.

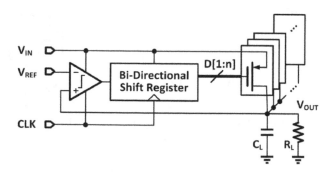

FIGURE 2.13
Schematic of a conventional digital LDO regulator.

The transient response time of the digital LDO regulator is proportional to its clock frequency and the size of each unit power transistor. Thus, coarse-fine tuning and adaptive clock techniques can be used to improve its transient response without increasing the standby power [32, 33]. In such a configuration, a slow clock will be used in steady state for low power, and a fast clock will be switched on for fast recovery when load transient is detected. In addition, digital codes can be easily reprogrammed to achieve different step transitions and/or DVS, which enable convenient system-level reconfigurations. However, because the power switches are fully turned on and operate in the linear region, digital LDO regulators suffer from very low PSRR, especially when their clock frequency is low. Digital LDO regulators also suffer from the problem of limit cycle oscillation (LCO) due to the finite resolution of the quantized output current [34]. When LCO occurs, the code $D[1:n]$ will periodically jump between two or multiple consecutive states, and results in undesirable steady-state output voltage ripples. Furthermore, a light load condition results in a larger load resistance R_L; then the same unit current will generate a larger LCO amplitude. Meanwhile, lower clock frequency results in longer charging and discharging times of C_L, which will also incite a larger LCO amplitude. On the other hand, a faster clock frequency will induce more oscillation steps of the LCO. This problem can be mitigated by using a multibit ADC, or by adding a feed-forward path with 1-bit strength to compensate for the slow main shift-register loop [35], or by introducing a dead zone to the comparator as discussed in [34].

2.4 Switched-Capacitor DC-DC Converter

The basic operation principle of the SC DC-DC converter is to step up or step down the power supply voltage by stacking and reconfiguring the capacitors periodically with multiple switches. Figure 2.14 shows the simplest SC converter with a voltage conversion ratio (VCR) of 1/2 (with current going from left to right) or 2 (with current going from right to left). The energy is transferred through the so-called flying capacitors C_{FLY} to the load capacitor C_L.

2.4.1 Charge Redistribution Loss

According to the charge balance law, which says that in a system of capacitors the sum of all charges leaving a node at any instance of charge transfer is equal to zero, the total initial charge equals the total final charge. However, when the charge is transferred between capacitors, a charge redistribution loss occurs [36, 37]. This can be easily understood by the analogy of pouring water (charge) from one cup (capacitor) to another cup (capacitor). Although the total amount of water remains unchanged, the potential of the water has been lost. In addition, the charge redistribution loss is proportional to $C\Delta V^2$,

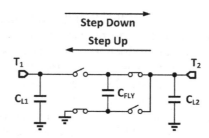

FIGURE 2.14
SC DC-DC converter with only one flying capacitor.

which means that the charge redistribution loss increases exponentially with the voltage difference between the capacitors that transferred energy.

Although the energy transfer between a capacitor and an inductor could be lossless, and the inductor-based DC-DC has a higher theoretical power conversion efficiency, as mentioned in the introduction part, the SC DC-DC converter is more suitable for low-power applications compared with its inductive counterpart. The required capacitor value shrinks with its output power, while the required inductor value of the inductor-based DC-DC goes the opposite way.

2.4.2 Switching Loss and Output Control

For SC DC-DC converters, their ideal output voltage $V_{\text{OUT,Ideal}}$ is equal to VCR × V_{IN}. However, the actual V_{OUT} will never be equal to that value as long as the converter is delivering current to the load. Even with a near-complete charge transfer every cycle, there will be a steady voltage drop on V_{OUT}, and the drop is proportional to $I_L/f_S C_{\text{FLY}}$, where I_L is the load current and f_S is the switching frequency. In addition, the charge transfer process will be affected by the RC time constant and the switching period, where the resistance R comes from the turn-on resistance of the switches. Therefore, for closed-loop load regulation, there are three common ways to regulate V_{OUT}: (1) tune the f_S [13, 38–40], (2) slice the SC power stage into multiple subcells and adjust the number of cells [41], and (3) digitally control the switch sizes or modulate the V_{GS} of the switches to slow down the charge transfer process [42–45].

As we know, the switching loss of the switches is proportional to $f_S C_{\text{SW}} V_{\text{DD}}^2$, where C_{SW} emerges from the parasitic capacitors of the switches. Therefore, reducing the f_S or the switch sizes (C_{SW}) would have the same effects on improving the efficiency. When f_S is reduced for lower power levels, the requirement of the charge transfer RC time constant is relaxed; then the switch sizes can also be reduced. Thus, when both f_S and the switch sizes are modulated, the efficiency can be further improved.

2.4.3 Ripple Reduction

The switching converters will unavoidably introduce output ripples. It should be noted that, from the energy point of view, the minimum V_{OUT} is the most important value for supply voltages. $V_{OUT,MIN}$ guarantees that the load can function correctly, while the part of the supply voltage higher than $V_{OUT,MIN}$ is wasted. The wasted power approximately equals $0.5(V_{OUT,MAX} - V_{OUT,MIN}) \times I_{OUT} = 0.5V_{RIPPLE} \times I_{OUT}$.

Figure 2.15 shows the interleaving multiphase architecture that is widely adopted for DC-DC conversion to achieve both input current ripple and output voltage ripple reduction [13, 46, 47]. The multiphase topology enables the energy to be delivered to the load in smaller packages, while maintaining the same switching frequency f_S. Consequently, it reduces the ripples without increasing the switching loss. The interleaving phases can be easily generated by a ring oscillator, which only adds negligible power and area overheads. Then the interleaving architecture saves area because it reduces both the input current ripple and the output voltage ripple, relaxing the requirements for both the input and output decoupling capacitors.

2.4.4 Parasitic Loss Reduction

Besides the charge redistribution loss and the switching loss, another major loss of the SC DC-DC converter is the parasitic loss, which derives from the parasitic capacitors of the flying capacitors. Especially for the fully-integrated solutions, the bottom-plate parasitic capacitor could be about 5% to 10% of the flying capacitor, while the top-plate parasitic capacitor is smaller (<1%).

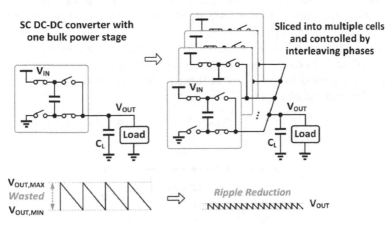

FIGURE 2.15
SC DC-DC converter with multiple interleaving phases for ripple reduction.

Here, for the bottom plate and the top plate, we mean the physical top and bottom plates, not the electrical positive and negative plates.

Figure 2.16 shows the SC converter with parasitic capacitors. The power losses associated with these parasitic capacitors are equal to $f_S(C_{P1} + C_{P2})V_{SW}^2$, where V_{SW} is the voltage swing of the parasitic capacitors. Since f_S is determined by the load demand, we will not change f_S for parasitic loss reduction. Thus, we can play with the two remaining terms: C_P and V_{SW}.

The MOS capacitors provide the highest capacitance density in standard CMOS technologies, being the most popular capacitor type for the SC DC-DC converters. Their bottom-plate parasitic is determined by the junction capacitors between the channel and the well or between the well and the substrate. One way to reduce the effective junction capacitance is to reduce the effective charging and discharging current of the parasitic capacitor by adding a large resistor on the DC-biasing path [48]; another way is to increase the reverse bias voltage of the PN junction for reducing the equivalent junction capacitance [49]. Both ways can be used together to further reduce the effective parasitic capacitance [50].

V_{SW} is related to the converter topologies. For the same conversion ratio, the voltage swings can be different by using summation or subtraction methods of stacking the flying capacitors [49]. Or, the charge on the parasitic capacitors can be recycled by introducing one or more clock phases in the conversion period [51–54]. The basic idea is to shorten the bottom plates of two or more time-interleaved flying capacitors during some intermediate clock phases, and let the charge on one parasitic capacitor move to another parasitic capacitor, recycling the charge and reducing the parasitic losses.

It is worth noting that the parasitic capacitor will definitely degrade the power conversion efficiency, but it does not necessarily deteriorate the output voltage [55]. For the C_{P1} in Figure 2.16, it gets the charge from the input and delivers the charge to the output, while C_{P2} only dumps charge to the ground. Therefore, C_{P1} participates in the power conversion process, but its conversion efficiency is low because the voltage variation on the capacitor is large. Further, the charge redistribution loss is proportional to $C\Delta V^2$. Therefore, it is suggested to make the bottom-plate parasitic capacitor C_{P1}, as in Figure 2.16, instead of C_{P2}.

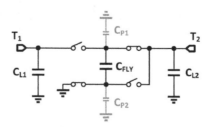

FIGURE 2.16
SC DC-DC converter with parasitic capacitors.

2.4.5 Realizing More VCRs

In low-power applications, the switching frequency is usually low. Therefore, the main losses are the charge redistribution loss and the parasitic loss. To reduce the charge redistribution loss, the actual output voltage should be close to the ideal output voltage. To achieve this within wide input and output ranges, it is highly desirable to have more VCRs.

With increased circuit complexity, more VCRs can be realized by cascading multiple SC converters and reconfiguring them according to the input and output voltages [56–58], or by utilizing more clock phases (not the interleaving phase) in one cycle [43, 49]. One more clock phase means one more degree of design freedom; thus, more VCR results can be resolved. However, power density is basically sacrificed with these schemes.

2.5 Summary and Future Works

This chapter discusses the possible ultralow standby power fully integrated voltage regulator solutions for low-duty-cycle IoT applications. First, we investigated the load conditions and system requirements of IoT devices. Some key design features for ultra-low-power IoT power management circuits can be summarized and listed as follows. (1) The load sleeps for most of the time and only operates at a very low duty cycle. (2) The load mainly operates in the near-threshold region that adds large load capacitors to the voltage regulator. (3) A large capacitive load needs to be supplied by a robust and stable voltage regulator. (4) An ultralow standby current is required for both the voltage regulator and the load. (5) Fast transition from the sleep mode to the active mode requires the regulator output voltage to adapt swiftly.

Meanwhile, we proposed a power management scheme based on the energy perspective. To support it, we designed an LDO regulator with switched biasing and -20 dB worst-case full-spectrum PSRR and 11 V/μs reference tracking speed. The digitally assisted analog LDO regulator can provide superior performances in both active and sleep modes, and can switch between the modes smoothly.

In addition, we introduced and discussed low-power circuit techniques and design considerations for both the LDO regulators and the SC DC-DC converters. The loss mechanisms of the SC converters were also analyzed in detail for improving the power conversion efficiency.

For future works, PSRR at low-power mode is still challenging for LDO regulators because the loop bandwidth is quite limited. Meanwhile, the digital LDO regulator can hardly provide PSRR because the power switches are fully turned on and operate in the linear region. For an SC DC-DC converter, realizing more VCRs with higher power density would be favorable for miniaturized devices. Of course, codesign with the IoT system will allow more degrees of freedom in the design, enabling creative breakthroughs, thus leading to near-optimum solutions.

References

1. Ki, W.-H., Lu, Y., Su, F., Tsui, C.-Y. 2012. Analysis and design strategy of on-chip charge pumps for micro-power energy harvesting applications. In *VLSI-SoC: Advanced Research for Systems on Chip*, Springer, Berlin, pp. 158–186. doi: 10.1007/978-3-642-32770-4_10.
2. Lu, Y. 2016. Digitally assisted low dropout regulator design for low duty cycle IoT applications. In *IEEE Asia Pacific Conference on Circuits and Systems (APCCAS)*, pp. 33–36. doi: 10.1109/ APCCAS.2016.7803888.
3. Wang, A., Chandrakasan, A. 2005. A 180-mV subthreshold FFT processor using a minimum energy design methodology. *IEEE Journal of Solid-State Circuits* 40, 310–319. doi: 10.1109/ JSSC.2004.837945.
4. Ramadass, Y.K., Chandrakasan, A.P. 2008. Minimum energy tracking loop with embedded DC–DC converter enabling ultra-low-voltage operation down to 250 mV in 65 nm CMOS. *IEEE Journal of Solid-State Circuits* 43, 256–265. doi: 10.1109/JSSC.2007.914720.
5. Burd, T.D., Pering, T.A., Stratakos, A.J., Brodersen, R.W. 2000. A dynamic voltage scaled microprocessor system. *IEEE Journal of Solid-State Circuits* 35, 1571–1580. doi: 10.1109/4.881202.
6. Alioto, M., Consoli, E., Rabaey, J.M. 2013. EChO reconfigurable power management unit for energy reduction in sleep-active transitions. *IEEE Journal of Solid-State Circuits* 48, 1921–1932. doi: 10.1109/JSSC.2013.2258816.
7. Hirairi, K., Okuma, Y., Fuketa, H., Yasufuku, T., Takamiya, M., Nomura, M., Shinohara, H., Sakurai, T. 2012. 13% power reduction in 16b integer unit in 40 nm CMOS by adaptive power supply voltage control with parity-based error prediction and detection (PEPD) and fully integrated digital LDO. In *IEEE International Solid-State Circuits Conference Digest of Technical Papers (ISSCC)*, pp. 486–488. doi: 10.1109/ISSCC.2012.6177102.
8. Wu, P.Y., Tsui, S.Y.S., Mok, P.K.T. 2010. Area- and power-efficient monolithic buck converters with pseudo-type III compensation. *IEEE Journal of Solid-State Circuits* 45, 1446–1455. doi: 10.1109/JSSC.2010.2047451.
9. Kim, W., Brooks, D.M., Wei, G.-Y. 2011. A fully-integrated 3-level DC/DC converter for nanosecond-scale DVS with fast shunt regulation. In *IEEE International Solid-State Circuits Conference Digest of Technical Papers (ISSCC)*, pp. 268–270. doi: 10.1109/ISSCC.2011.5746313.
10. Jain, R., Geuskens, B.M., Kim, S.T., Khellah, M.M., Kulkarni, J., Tschanz, J.W., De, V. 2014. A 0.45–1 V fully-integrated distributed switched capacitor DC-DC converter with high density MIM capacitor in 22 nm Tri-Gate CMOS. *IEEE Journal of Solid-State Circuits* 49, 917–927. doi: 10.1109/JSSC.2013.2297402.
11. Kim, S.T., Shih, Y.-C., Mazumdar, K., Jain, R., Ryan, J.F., Tokunaga, C., Augustine, C., Kulkarni, J.P., Ravichandran, K., Tschanz, J.W., Khellah, M.M., De, V. 2016. Enabling wide autonomous DVFS in a 22 nm graphics execution core using a digitally controlled fully integrated voltage regulator. *IEEE Journal of Solid-State Circuits* 51, 18–30. doi: 10.1109/JSSC.2015.2457920.
12. Cheng, L., Liu, Y., Ki, W.-H. 2014. A 10/30 MHz fast reference-tracking buck converter with DDA-based type-III compensator. *IEEE Journal of Solid-State Circuits* 49, 2788–2799. doi: 10.1109/JSSC.2014.2346770.

13. Lu, Y., Jiang, J., Ki, W.H. 2017. A multiphase switched-capacitor DC–DC converter ring with fast transient response and small ripple. *IEEE Journal of Solid-State Circuits* 52, 579–591. doi: 10.1109/JSSC.2016.2617315.

14. Lu, Y., Ki, W.-H., Yue, C.P. 2014. A 0.65 ns-response-time 3.01 ps FOM fully-integrated low-dropout regulator with full-spectrum power-supply-rejection for wideband communication systems. In *2014 IEEE International Solid-State Circuits Conference Digest of Technical Papers (ISSCC)*, pp. 306–307. doi: 10.1109/ISSCC.2014.6757446.

15. Lu, Y., Wang, Y., Pan, Q., Ki, W.-H., Yue, C.P. 2015. A fully-integrated low-dropout regulator with full-spectrum power supply rejection. *IEEE Transactions on Circuits and Systems I: Regular Papers* 62, 707–716. doi: 10.1109/TCSI.2014.2380644.

16. Lu, Y., Li, C., Zhu, Y., Huang, M., U, S.-P., Martins, R.P. 2016. A 312 ps response-time LDO with enhanced super source follower in 28 nm CMOS. *Electronics Letters* 52, 1368–1370. doi: 10.1049/el.2016.1719.

17. Leung, K.N., Mok, P.K.T. 2003. A capacitor-free CMOS low-dropout regulator with damping-factor-control frequency compensation. *IEEE Journal of Solid-State Circuits* 38, 1691–1702. doi: 10.1109/JSSC.2003.817256.

18. Milliken, R.J., Silva-Martinez, J., Sanchez-Sinencio, E. 2007. Full on-chip CMOS low-dropout voltage regulator. *IEEE Transactions on Circuits and Systems I: Regular Papers* 54, 1879–1890. doi: 10.1109/TCSI.2007.902615.

19. Man, T.Y., Leung, K.N., Leung, C.Y., Mok, P.K.T., Chan, M. 2008. Development of single-transistor-control LDO based on flipped voltage follower for SoC. *IEEE Transactions on Circuits and Systems I: Regular Papers* 55, 1392–1401. doi: 10.1109/TCSI.2008.916568.

20. Guo, J., Leung, K.N. 2010. A 6-μW chip-area-efficient output-capacitorless LDO in 90-nm CMOS technology. *IEEE Journal of Solid-State Circuits* 45, 1896–1905. doi: 10.1109/JSSC.2010.2053859.

21. Tan, X.L., Chong, S.S., Chan, P.K., Dasgupta, U. 2014. A LDO regulator with weighted current feedback technique for 0.47 nF–10 nF capacitive load. *IEEE Journal of Solid-State Circuits* 49, 2658–2672. doi: 10.1109/JSSC.2014.2346762.

22. Rincon-Mora, G., Allen, P.E. 1998. A low-voltage, low quiescent current, low drop-out regulator. *IEEE Journal of Solid-State Circuits* 33, 36–44. doi: 10.1109/4.654935.

23. Lam, Y.-H., Ki, W.-H., Tsui, C.-Y. 2006. Adaptively-biased capacitor-less CMOS low dropout regulator with direct current feedback. In *Asia and South Pacific Conference on Design Automation*, pp. 104–105. doi: 10.1109/ASPDAC.2006.1594658.

24. Zhan, C., Ki, W.-H. 2010. Output-capacitor-free adaptively biased low-dropout regulator for system-on-chips. *IEEE Transactions on Circuits and Systems I: Regular Papers* 57, 1017–1028. doi: 10.1109/TCSI.2010.2046204.

25. Gupta, V., Rincon-Mora, G., Raha, P. 2004. Analysis and design of monolithic, high PSR, linear regulators for SoC applications. In *Proceedings of IEEE International SOC Conference*, pp. 311–315. doi: 10.1109/SOCC.2004.1362447.

26. den Besten, G.W., Nauta, B. 1998. Embedded 5 V-to-3.3 V voltage regulator for supplying digital IC's in 3.3 V CMOS technology. *IEEE Journal of Solid-State Circuits* 33, 956–962. doi: 10.1109/4.701230.

27. Carvajal, R.G., Ramirez-Angulo, J., Lopez-Martin, A., Torralba, A., Galan, J.A., Carlosena, A., Chavero, F.M. 2005. The flipped voltage follower: A useful cell for low-voltage low-power circuit design. *IEEE Transactions on Circuits and Systems I: Regular Papers* 52, 1276–1291. doi: 10.1109/TCSI.2005.851387.

28. Okuma, Y., Ishida, K., Ryu, Y., Zhang, X., Chen, P.-H., Watanabe, K., Takamiya, M., Sakurai, T. 2010. 0.5-V input digital LDO with 98.7% current efficiency and 2.7-μA quiescent current in 65 nm CMOS. In *2010 IEEE Custom Integrated Circuits Conference (CICC)*, pp. 1–4. doi: 10.1109/CICC.2010.5617586

29. Lee, Y.-H., Peng, S.-Y., Chiu, C.-C., Wu, A.C.-H., Chen, K.-H., Lin, Y.-H., Wang, S.-W., Tsai, T.-Y., Huang, C.-C., Lee, C.-C. 2013. A low quiescent current asynchronous digital-LDO with PLL-modulated fast-DVS power management in 40 nm SoC for MIPS performance improvement. *IEEE Journal of Solid-State Circuits* 48, 1018–1030. doi: 10.1109/JSSC.2013.2237991.

30. Gangopadhyay, S., Somasekhar, D., Tschanz, J.W., Raychowdhury, A. 2014. A 32 nm embedded, fully-digital, phase-locked low dropout regulator for fine grained power management in digital circuits. *IEEE Journal of Solid-State Circuits* 49, 2684–2693. doi: 10.1109/JSSC.2014.2353798.

31. Nasir, S.B., Gangopadhyay, S., Raychowdhury, A. 2016. All-digital low-dropout regulator with adaptive control and reduced dynamic stability for digital load circuits. *IEEE Transactions on Power Electronics* 31, 8293–8302. doi: 10.1109/TPEL.2016.2519446.

32. Huang, M., Lu, Y., Sin, S.W., U, S.-P., Martins, R.P. 2016. A fully integrated digital LDO with coarse–fine-tuning and burst-mode operation. *IEEE Transactions on Circuits and Systems II: Express Briefs* 63, 683–687. doi: 10.1109/TCSII.2016.2530094.

33. Huang, M., Lu, Y., U, S.-P., Martins, R.P. 2017. An output-capacitor-free analog-assisted digital low-dropout regulator with tri-loop control. In *2017 IEEE International Solid-State Circuits Conference (ISSCC)*, pp. 342–343. doi: 10.1109/ISSCC.2017.7870401.

34. Huang, M., Lu, Y., Sin, S.W., U, S.-P., Martins, R.P., Ki, W.H. 2016. Limit cycle oscillation reduction for digital low dropout regulators. *IEEE Transactions on Circuits and Systems II: Express Briefs* 63, 903–907. doi: 10.1109/TCSII.2016.2534778.

35. Huang, M., Lu, Y., Seng-Pan, U., Martins, R.P. 2016. A digital LDO with transient enhancement and limit cycle oscillation reduction. In *2016 IEEE Asia Pacific Conference on Circuits and Systems (APCCAS)*, pp. 25–28. doi: 10.1109/APCCAS.2016.7803886.

36. Ki, W.-H., Su, F., Tsui, C. 2005. Charge redistribution loss consideration in optimal charge pump design. In *IEEE International Symposium on Circuits and Systems (ISCAS)*, vol. 2, pp. 1895–1898. doi: 10.1109/ISCAS.2005.1464982.

37. Ki, W.-H., Lu, Y., Su, F., Tsui, C.-Y. 2011. Design and analysis of on-chip charge pumps for micro-power energy harvesting applications. In *2011 IEEE/IFIP 19th International Conference on VLSI and System-on-Chip (VLSI-SoC)*, pp. 374–379. doi: 10.1109/VLSISoC.2011.6081612.

38. Ramadass, Y.K., Fayed, A.A., Chandrakasan, A.P. 2010. A fully-integrated switched-capacitor step-down DC-DC converter with digital capacitance modulation in 45 nm CMOS. *IEEE Journal of Solid-State Circuits* 45, 2557–2565. doi: 10.1109/JSSC.2010.2076550.

39. Lu, Y., Jiang, J., Ki, W.-H., Yue, C.P., Sin, S.-W., Seng-Pan, U., Martins, R.P. 2015. A 123-phase DC-DC converter-ring with fast-DVS for microprocessors. In *2015 IEEE International Solid-State Circuits Conference (ISSCC)*, pp. 1–3. doi: 10.1109/ISSCC.2015.7063077.

40. Jiang, J., Lu, Y., Ki, W.H., U, S.-P., Martins, R.P. 2017. A dual-symmetrical-output switched-capacitor converter with dynamic power cells and minimized cross regulation for application processors in 28 nm CMOS. In *2017 IEEE International Solid-State Circuits Conference (ISSCC)*, pp. 344–345. doi: 10.1109/ISSCC.2017.7870402.

41. Lu, Y., Ki, W.H., Patrick Yue, C. 2016. An NMOS-LDO regulated switched-capacitor DC–DC converter with fast-response adaptive-phase digital control. *IEEE Transactions on Power Electronics* 31, 1294–1303. doi: 10.1109/TPEL.2015.2420572.

42. Jiang, J., Lu, Y., Ki, W.H. 2016. A digitally-controlled 2-/3-phase 6-ratio switched-capacitor DC-DC converter with adaptive ripple reduction and efficiency improvements. In *42nd European Solid-State Circuits Conference*, pp. 441–444. doi: 10.1109/ESSCIRC.2016.7598336.

43. Jiang, J., Ki, W.H., Lu, Y. 2017. Digital 2-/3-phase switched-capacitor converter with ripple reduction and efficiency improvement. *IEEE Journal of Solid-State Circuits* 52, 1836–1848. doi: 10.1109/JSSC.2017.2679065.

44. Lee, H., Mok, P.K.T. 2007. An SC voltage doubler with pseudo-continuous output regulation using a three-stage switchable opamp. *IEEE Journal of Solid-State Circuits* 42, 1216–1229. doi: 10.1109/JSSC.2007.897133.

45. Su, F., Ki, W.-H., Tsui, C. 2009. Regulated switched-capacitor doubler with interleaving control for continuous output regulation. *IEEE Journal of Solid-State Circuits* 44, 1112–1120. doi: 10.1109/JSSC.2009.2014727.

46. Le, H.-P., Sanders, S.R., Alon, E. 2011. Design techniques for fully integrated switched-capacitor DC-DC converters. *IEEE Journal of Solid-State Circuits* 46, 2120–2131. doi: 10.1109/JSSC.2011.2159054.

47. Pique, G.V. 2012. A 41-phase switched-capacitor power converter with 3.8 mV output ripple and 81% efficiency in baseline 90nm CMOS. In *2012 IEEE International Solid-State Circuits Conference Digest of Technical Papers (ISSCC)*, pp. 98–100. doi: 10.1109/ISSCC.2012.6176892.

48. Le, H.-P., Crossley, J., Sanders, S.R., Alon, E. 2013. A sub-ns response fully integrated battery-connected switched-capacitor voltage regulator delivering 0.19W/mm² at 73% efficiency. In *2013 IEEE International Solid-State Circuits Conference Digest of Technical Papers (ISSCC)*, pp. 372–373. doi: 10.1109/ISSCC.2013.6487775.

49. Jiang, J., Lu, Y., Huang, C., Ki, W.-H., Mok, P.K.T. 2015. A 2-/3-phase fully integrated switched-capacitor DC-DC converter in bulk CMOS for energy-efficient digital circuits with 14% efficiency improvement. In *2015 IEEE International Solid-State Circuits Conference (ISSCC)*, pp. 1–3. doi: 10.1109/ISSCC.2015.7y*063078.

50. Butzen, N., Steyaert, M. 2017. 10.1 A 1.1W/mm²-power-density 82%-efficiency fully integrated 3:1 switched-capacitor DC-DC converter in baseline 28 nm CMOS using stage outphasing and multiphase soft-charging. In *2017 IEEE International Solid-State Circuits Conference (ISSCC)*, pp. 178–179. doi: 10.1109/ISSCC.2017.7870319.

51. Su, F. 2008. Topology, control and implementation of switched-capacitor DC-DC power converters for portable applications. PhD thesis, Hong Kong University of Science and Technology. doi: 10.14711/thesis-b1029348.

52. Tong, T., Zhang, X., Kim, W., Brooks, D., Wei, G.Y. 2013. A fully integrated battery-connected switched-capacitor 4:1 voltage regulator with 70% peak efficiency using bottom-plate charge recycling. In *Proceedings of the IEEE 2013 Custom Integrated Circuits Conference*, pp. 1–4. doi: 10.1109/CICC.2013.6658485.
53. Lisboa, P.C., Pérez-Nicoli, P., Veirano, F., Silveira, F. 2016. General top/bottom-plate charge recycling technique for integrated switched capacitor DC-DC converters. *IEEE Transactions on Circuits and Systems I: Regular Papers* 63, 470–481. doi: 10.1109/TCSI.2016.2528478.
54. Butzen, N., Steyaert, M.S.J. 2016. Scalable parasitic charge redistribution: Design of high-efficiency fully integrated switched-capacitor DC–DC converters. *IEEE Journal of Solid-State Circuits* 51, 2843–2853. doi: 10.1109/JSSC.2016.2608349.
55. Meyvaert, H., Van Breussegem, T., Steyaert, M. 2013. A 1.65 W fully integrated 90 nm bulk cmos capacitive DC-DC converter with intrinsic charge recycling. *IEEE Transactions on Power Electronics* 28, 4327–4334. doi: 10.1109/TPEL.2012.2230339.
56. Bang, S., Wang, A., Giridhar, B., Blaauw, D., Sylvester, D. 2013. A fully integrated successive-approximation switched-capacitor DC-DC converter with 31 mV output voltage resolution. In *2013 IEEE International Solid-State Circuits Conference (ISSCC)*, pp. 370–371. doi: 10.1109/ISSCC.2013.6487774.
57. Salem, L.G., Mercier, P.P. 2014. A recursive switched-capacitor DC-DC converter achieving $2^N - 1$ ratios with high efficiency over a wide output voltage range. *IEEE Journal of Solid-State Circuits* 49, 2773–2787. doi: 10.1109/JSSC.2014.2353791.
58. Jung, W., Sylvester, D., Blaauw, D. 2016. 12.1 A rational-conversion-ratio switched-capacitor DC-DC converter using negative-output feedback. In *2016 IEEE International Solid-State Circuits Conference (ISSCC)*, pp. 218–219. doi: 10.1109/ISSCC.2016.7417985.

3

On-Chip Regulators for Low-Voltage and Portable Systems-on-Chip

Emre Salman

CONTENTS

3.1 Introduction

Efficient voltage regulation and conversion are essential mechanisms in modern integrated circuit (IC) design process due to power management and heterogeneous computing [1]. Specifically, fully monolithic on-chip voltage regulation has emerged as a critical process for a variety of low-power design methodologies, such as voltage islands (ranging from ultralow voltages in the range of 0.4–0.5 V to higher voltages in the range of 1.2–1.4 V), dynamic voltage (and frequency) scaling, low-voltage clocking, and near-threshold computing [2–4]. Furthermore, on-chip voltage regulators play a critical role to ensure sufficient power integrity since it is highly challenging to maintain power supply noise within a tolerable range when the supply voltage is low and the load current is high [5–7]. Power supply noise not only affects the timing characteristics within synchronous digital circuits, but also degrades the overall signal integrity in analog and mixed-signal circuits [8]. For example, a fully integrated voltage regulator (FIVR) was developed for the Intel Haswell microarchitecture, allowing dynamically managed multiple power domains [9].

The rest of the chapter is organized as follows. Opportunities provided by monolithic voltage regulation and related challenges are summarized in the following subsections. A broad overview of primary voltage regulator topologies is provided in Section 3.2 with emphasis on low-dropout (LDO) regulators, switched-capacitor-based regulators, and switching buck regulators. A fully monolithic hybrid regulator topology is described in Section 3.3 with application to low-voltage systems such as near-threshold computing. Finally, the chapter is summarized in Section 3.4.

3.1.1 Opportunities Provided by Monolithic Regulators

On-chip integration of a voltage regulator on the same die as the load circuit, as illustrated in Figure 3.1(b), has several advantages compared with an off-chip voltage regulator, as illustrated in Figure 3.1(a). These advantages include

- Reduction in conduction loss due to reduced parasitic interconnect impedances
- Superior voltage regulation characteristics
- Enhanced power supply noise characteristics
- Reduced number of power pads and less metal resources for multi-voltage systems

In off-chip regulators, the parasitic impedances of the interconnect among the devices, pads, and package dissipate significant energy, thereby reducing the overall efficiency of a regulator. Integrating a voltage regulator with the load circuit can potentially reduce these parasitic losses since the interconnect length is significantly shorter [10,11].

Furthermore, a monolithic regulator outperforms an off-chip regulator in terms of regulation characteristics since the regulator is physically closer to the load, producing a faster response. Thus, the regulator exhibits reduced sensitivity to changes in the load current. The power supply noise caused by the parasitic interconnect impedances and package inductance is also reduced.

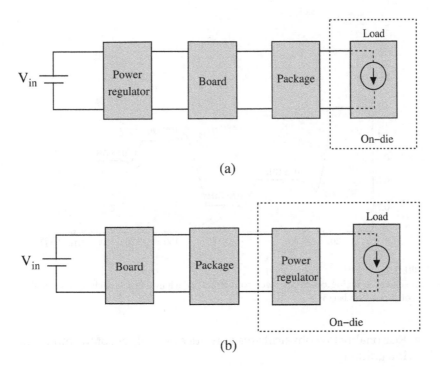

FIGURE 3.1
Conceptual representation of an (a) off-chip and an (b) on-chip power converter.

Another advantage of on-chip regulators for multivoltage systems (voltage islands) is the reduction in the number of power pads. The metal resources allocated for the global power grids are also reduced since a separate global power distribution network for each voltage is not required. Similarly, on-chip regulators are considered to be an enabling technology for dynamic voltage frequency scaling (DVFS), where the power supply voltage is temporarily adjusted based on the required computation [2,12]. This technique requires fast voltage transients, on the scale of nanoseconds, which is possible with the use of on-chip power regulators. The timing diagram of a fine-grain DVFS scheme is illustrated in Figure 3.2, where four different voltage and frequency levels are achieved using on-chip conversion [13].

3.1.2 Challenges in Designing Monolithic Regulators

Traditional trade-offs in regulator design are exacerbated when all the components are required to be on-chip. Specifically, it is highly challenging to simultaneously satisfy the following design objectives:

- Sufficiently high energy efficiency (in the range of 80%–90%) to minimize power loss during regulation and conversion, while providing the required output current

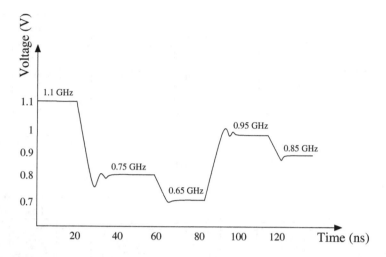

FIGURE 3.2
DVFS scheme utilizing on-chip power converters to achieve fast switching transients in the range of nanoseconds [13].

- Reasonably low physical area to reduce overall cost of the monolithic integration
- Enhanced regulation characteristics to ensure sufficiently low output voltage ripple during voltage conversion
- Sufficient thermal integrity since on-chip regulators that provide high output current are likely to cause thermal hot spots, thereby increasing the cooling cost

Linear regulators typically satisfy the area requirement and offer a low-cost solution, but fail to achieve the energy efficiency constraint. Alternatively, switching buck regulators achieve high energy efficiency due to ideally lossless circuit elements, i.e., the capacitor and inductor. Switching regulators, however, consume significant area due to these passive elements within the LC filter, particularly the inductor. Switched-capacitor-based regulators have received significant attention to simultaneously achieve the aforementioned four design objectives. An important challenge for switched-capacitor-based regulators is the difficulty of adjusting output voltage according to application requirement, i.e., achieving variable voltage conversion ratios. The primary characteristics of existing voltage regulators are described in the following section.

3.2 Overview of Primary Voltage Regulator Topologies

There are primarily three types of voltage regulators: (1) linear converters such as LDO regulators, (2) switched-capacitor-based DC-DC regulators, and (3) switching buck regulators. These topologies are discussed in the following subsections. A qualitative comparison of these topologies is also provided in the last subsection.

3.2.1 Low-Dropout Regulators

LDO voltage regulators are a type of linear DC-DC converter where the power efficiency is enhanced by lowering the voltage drop across the pass transistor, i.e., between the input and output of the regulator. This improvement is achieved by replacing the *common drain* structure with a *common source* topology [14–17]. The voltage drop $V_{drop-out}$ is

$$V_{drop-out} = I_{in} \times R_{on}, \tag{3.1}$$

where R_{on} is the on-channel resistance of the power transistor. The power dissipation is reduced due to the smaller voltage drop, making LDO voltage regulators a suitable candidate for low-voltage, low-power applications. LDO regulators can also be used for isolating input power supply noise in noise-sensitive applications [18].

As depicted in Figure 3.3, a conventional LDO regulator is composed of an error amplifier, reference voltage generator, power transistor with a common source configuration, passive resistors to achieve voltage division, and output capacitance (or decoupling capacitance) to satisfy stability constraints. Resistors R_1 and R_2 are added since the output voltage of an LDO converter is at the drain rather than the source terminal of the

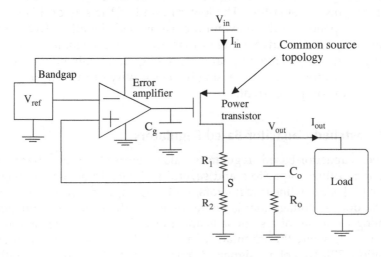

FIGURE 3.3
Block diagram of an LDO voltage regulator consisting of an error amplifier, reference voltage generator, power transistor with a common source configuration, passive resistors, and output capacitance.

power transistor. Any variation in the output voltage is sensed at node S and compared with V_{ref} by the error amplifier. The current flowing through the pass transistor is accordingly adjusted to maintain a constant output voltage. Ideally, the output voltage is maintained at $(V_{ref}/R_2) \times (R_1 + R_2)$. The temperature coefficient of the regulator is determined by the temperature dependence of the reference voltage generator and the input offset voltage of the error amplifier [18].

The design process of an LDO regulator exhibits several challenging tradeoffs. For example, the *quiescent current* of the regulator plays an important role in determining the overall power efficiency. The quiescent current refers to the input current of the regulator when there is no output (load and decoupling) capacitance. The effect of the quiescent current on the current efficiency becomes significant particularly when the load current is lower. Alternatively, for those applications where the output load current is significantly high the majority of the time, the energy efficiency is primarily determined by the ratio of the output voltage to the input voltage. In this case, LDO regulators are particularly advantageous when the voltage difference between the input and output is small. Smaller area and fast load regulation due to the small output impedance are also important advantages. However, since a higher load current is typically a temporary condition, the quiescent current has a vital effect on the overall power efficiency. Resistances R_1 and R_2 are adjusted to maintain a sufficiently low quiescent current, such as 1% of the load current [19]. A low quiescent current, however, typically degrades the transient response of the regulator, negatively affecting the regulation characteristics. The transient response time can be improved by increasing the slew rate at the output of the amplifier, which can be achieved by downsizing the power transistors. The lower bound of the size of this transistor, however, depends on the maximum current load. Finally, LDO regulators typically require a relatively large capacitor to ensure stable behavior, particularly when the open-loop gain is high. High loop gain enhances the regulation characteristics by decreasing the sensitivity of the output voltage to changes in the output current.

3.2.2 Switched-Capacitor-Based Regulators

Switched-capacitor-based regulators, also referred to as charge pump converters, utilize capacitors and several switches to achieve voltage conversion [20]. Unlike linear regulators, switched-capacitor converters can produce an output voltage that is higher or lower than the input voltage. The operating principle of a switched-capacitor DC-DC converter is illustrated in Figure 3.4, where the schematic of a *voltage doubler* is shown, without any regulation [20]. Two phase signals, ϕ_1 and ϕ_2, control the switches within the circuit. Note that these signals are out of phase to prevent any overlap. In the first phase, the phase 1 switches are closed and the phase 2 switches are open, thereby charging the capacitance C_1 to V_{in}, as depicted in Figure 3.5(a). In the

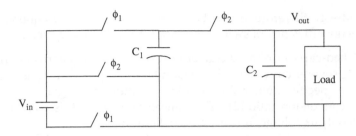

FIGURE 3.4
Schematic representation of a voltage doubler based on a switched-capacitor DC-DC converter.

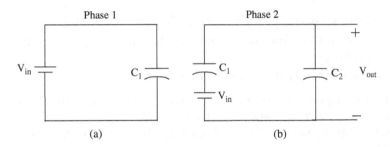

FIGURE 3.5
Operation of a switched-capacitor voltage doubler: (a) phase 1 and (b) phase 2.

second phase, as illustrated in Figure 3.5(b), the phase 2 switches are closed and the phase 1 switches are open. In this case, the input power supply voltage is connected in series with the capacitance C_1, which had been charged to V_{in} in the previous phase. This series connection produces a voltage of $2 \times V_{in}$ across the capacitance C_2, which is the output voltage applied across the load circuit. Note that the capacitance C_1 behaves as a charge pump during the second phase. Also note that during ϕ_1, the output voltage is maintained close to $2 \times V_{in}$, assuming that the switching frequency is sufficiently high. The minimum switching frequency is primarily determined by the load current characteristics and the size of C_1 and C_2. For example, the size of C_1 can be reduced if a higher switching frequency is used [21]. This dependence produces two trade-offs in the design of a switched-capacitor DC-DC converter:

- The switches are typically implemented using metal-oxide semiconductor (MOS) transistors, which are sized based on the switching frequency. The higher the switching frequency, the greater the width of these transistors. A trade-off therefore exists between the size of the capacitors and the width of the transistors operating as switches.

- As the size of the transistors increases, dynamic power dissipation also increases due to the larger gate oxide capacitance. Another trade-off therefore exists between the switching frequency and energy efficiency of switched-capacitor DC-DC converters.

Switched-capacitor-based converters can achieve any conversion ratio by cascading several converters. For example, the input voltage is first multiplied by a specific integer, and divided by another integer to produce the required conversion ratio [21]. Dynamically changing this conversion ratio to adjust output voltage, however, is challenging.

As illustrated in Figure 3.4, a typical switched-capacitor DC-DC converter does not use feedback to achieve load regulation. The three methods for achieving load regulation are

- To vary C_1 to compensate for changes in the output voltage
- To vary the conductance of the switches that charge/discharge the capacitors
- To vary the duty cycle of the switching period

The first option is limited by the energy efficiency since a lower C_1 reduces the power efficiency. Alternatively, the second and third options require energy-consuming feedback circuitry, which also degrades the power efficiency.

3.2.3 Switching Buck Regulators

A switching buck regulator is a stepdown DC-DC voltage regulator to supply power to various circuit modules, such as a CPU core, memory core, or accelerator module. A typical single-phase buck converter consists of (1) a switch network that generates an AC signal, (2) a second-order low-pass filter that passes the DC component of the AC signal to the output, and (3) a feedback path that regulates the output voltage by changing the duty cycle of the AC signal [1]. These primary elements of a buck regulator are shown in Figure 3.6 where the power transistors, cascaded powers, and a pulse width modulator to regulate the output voltage are depicted. Single- and multiphase operations of a switching buck regulator are described in the following subsection.

3.2.3.1 Single-Phase Operation

Typical design specifications of a switching buck converter include input and output voltages, power efficiency, load current, voltage ripple, and transient response. The low-pass filter, consisting of an inductor and capacitor, is a critical element within the buck converter since the output voltage characteristics depend on the quality of this filter. The parasitic effective series resistance (ESR) of the inductor plays an important role in the resistive loss and the overall performance of the buck converter. A larger inductance (required

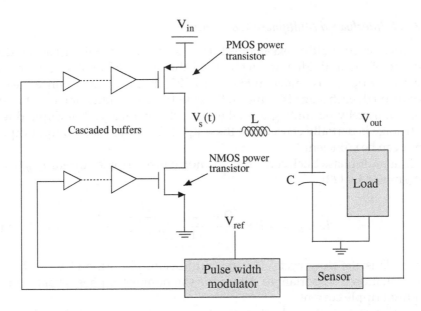

FIGURE 3.6
Schematic of a buck converter utilizing power transistors, cascaded buffers, and a pulse width modulator to regulate the output voltage.

to reduce ripple) typically produces a larger ESR, which in turn increases the resistive loss and causes a nonnegligible voltage drop at the output, particularly if the load current is sufficiently high.

For a single-phase buck converter, the required inductance can be determined by [1]

$$L = \frac{(V_{in} - V_{out})D}{2\Delta I_L f_s},$$ (3.2)

where V_{in} and V_{out} are, respectively, input and output voltages, D is duty cycle, ΔI_L is the current ripple (half of the peak-to-peak current), and f_s is switching frequency. Assuming the output voltage ripple cannot exceed 5% of the output voltage, the minimum required capacitor C_{out} is determined by [1]

$$C_{out} \geq \frac{5(V_{in} - V_{out})D}{4V_{out}Lf_s^2}.$$ (3.3)

Single-phase buck converters are sufficient for applications with low load current [22], but power dissipation and efficiency suffer at higher load currents. Thus, interleaved multiphase buck regulators have been considered for applications with high load current since peak ripple currents can be effectively reduced through this method [23, 24]. Interleaved multiphase buck regulators are discussed in the following subsection.

3.2.3.2 Interleaved Multiphase Operation

An interleaved multiphase architecture has been commonly used to reduce the size of the individual inductors (and therefore ESR) without increasing the output ripple, as shown in Figure 3.7 [25]. In this method, since the current through each stage is reduced, the constraint on inductor current is also relaxed, thereby permitting a smaller inductor per stage. The ripple due to each stage is partially canceled at the output. Thus, a smaller output capacitance can be sufficient.

In a multiphase buck converter, the normalized ripple current I_{Rip_norm} is determined by [22]

$$I_{Rip\_norm} = P \times \frac{[D - \frac{\lfloor m \rfloor}{P}] \times [\frac{1+\lfloor m \rfloor}{P} - D]}{(1 - D) \times D}, \tag{3.4}$$

where D is the duty cycle, P is the number of phases, and $m = D \times P$. This equation is important to determine the number of phases based on the required ripple current.

In this case, multiple buck converters operate in a parallel fashion with a 90° phase difference. Each regulator has an individual inductor, but shares the same output capacitor. Thus, the high ripple across each inductor is partially canceled at the output.

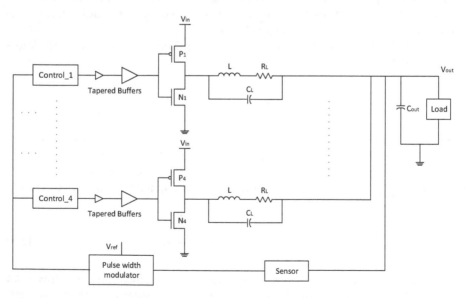

FIGURE 3.7
An interleaved multiphase switching buck regulator architecture.

3.2.4 Qualitative Comparison

LDO regulators are cost-effective and have a relatively fast transient response, but these regulators suffer from low power efficiency, less than 60% in most of the cases [26]. This limitation is exacerbated as the conversion ratio increases or output voltage decreases.

Switched-capacitor converters exhibit enhanced power efficiency at relatively small area, but suffer from poor regulation capability since the switching frequency should be modified to regulate the output [27]. This process is slow since a voltage-controlled oscillator is needed to vary the switching frequency, increasing the response time. Furthermore, it is challenging to dynamically tune the voltage conversion ratio.

Finally, switching buck converters can achieve high efficiency and large output current at the expense of a high-quality inductor [9]. Since integrating a high-quality inductor on chip is very costly, buck regulators typically consist of an external, discrete inductor. Another option is to utilize the flip-chip package for developing a package-embedded spiral inductor. For example, in [28], existing wirebond inductance of a standard package (instead of spiral metals) has been utilized for a buck converter. Similarly, in [29], both the wirebond and lead frame inductance have been engineered to be used with an integrated buck converter. These approaches reduce the overall cost (since an existing package structure is leveraged for inductance) at the expense of higher inductance variability and reduced flexibility for the value of inductance. Thus, additional mechanisms, such as extra calibration loops, are required to alleviate these challenges [28]. In [30], package-embedded inductors have been discussed with emphasis on building high-Q inductors within the routing layers of an organic package.

In [31, 32], a higher inductance with a reasonable quality factor was achieved by exploiting the greater flexibility of package area compared with die area. This package-embedded spiral inductor was connected to the die via low-resistance C4 bumps. Thus, the switching frequency of the buck regulator was reduced to minimize dynamic power loss and enhance power efficiency. Specifically, the switching frequency was 50 MHz, which is significantly smaller than typical switching frequencies used in buck regulators with on-chip inductors (477 MHz [11], 200 MHz [33], 170 MHz [34], and 300 MHz [35]). Furthermore, package flexibility can be further utilized to develop an *array of package-embedded spiral inductors* for an interleaved multiphase buck regulator architecture. Thus, power efficiency and output ripple can be further enhanced.

Existing work has also investigated the feasibility of increasing switching frequency to reduce the required inductance. In this case, however, the dynamic loss increases, thereby reducing the power efficiency [11]. For example, for inductance values in the low nanohenry range, the switching frequency should be increased to several hundreds of megahertz to obtain an acceptable current and voltage ripple at the output. The switching loss at these frequencies increases by two orders of magnitude compared with high-kilohertz or low-megahertz operating frequencies (assuming constant transistor sizes) [29].

3.3 Monolithic Hybrid Regulator Topology for Low-Voltage Applications

Near-threshold computing has received significant attention due to enhanced energy efficiency, particularly for mobile systems-on-a-chip (SoCs) [36]. Highly parallelized architectures based on near-threshold operation have been proposed as a possible solution to dark silicon [37]. Developing an integrated voltage regulator module with application to near-threshold operation is challenging due to low output voltages in the range of 0.5 V. The regulator should simultaneously satisfy high power efficiency and power density (to minimize area overhead). Furthermore, the output ripple should be minimized since near-threshold circuits are highly sensitive to power supply variations (due to near-exponential dependence).

Hybrid regulators have also been developed to exploit the advantages of both LDOs and switching converters [38,39], as conceptually depicted in Figure 3.8(a). Existing hybrid topologies, however, suffer in near-threshold

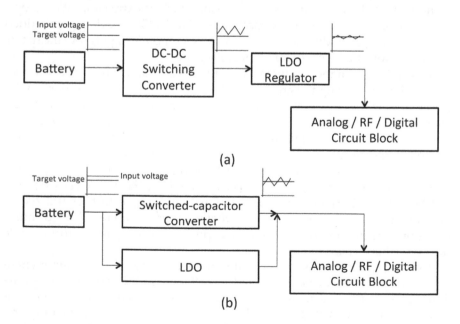

FIGURE 3.8
Conceptual block diagram of a (a) conventional hybrid regulator and a (b) proposed hybrid regulator.

operation, as further discussed in the following section. A novel hybrid topology, as shown in Figure 3.8(b), was developed [27]. This topology can produce low output voltages at high power efficiency. The proposed approach achieves approximately 85% power efficiency while supplying 100 mA output current at 0.5 V with a maximum ripple voltage of 22 mV.

3.3.1 Overview of Hybrid Regulator Topologies

In existing hybrid regulator approaches, a DC-DC switching converter is combined in series with an LDO, as shown in Figure 3.8(a) [38, 39]. The switched-capacitor circuit functions as a converter without any regulation capability, whereas the LDO regulates the output voltage without any conversion. Thus, the circuit enhances power efficiency since LDO has a near-unity voltage conversion ratio. Regulation is also enhanced due to an LDO with fast transient response at the output. Feed-forward ripple cancellation has also been proposed to further improve the regulation process [38]. This topology, however, suffers in near-threshold operation with large output current and low output voltage for three reasons:

- Power transistor of the LDO suffers from low $|V_{GS}|$ since the voltage conversion is achieved by the previous stage (switched-capacitor converter). This low input voltage makes it challenging to supply high current at the output.
- At high output current, the voltage drop across the power transistor (within an LDO) becomes nonnegligible, requiring a higher input voltage at the LDO. Higher input voltage, however, degrades the power efficiency.
- The maximum output load current cannot be larger than the current supplied by the switched-capacitor converter due to the series connection. Thus, the DC-DC switching converter needs to be optimized for the maximum load current rather than the nominal load current.

These limitations are exacerbated and the power efficiency is further degraded with reduced output voltages, as in near-threshold computing. Thus, a novel hybrid topology was proposed where the switched-capacitor converter and LDO operate in a parallel fashion, as conceptually illustrated in Figure 3.8(b). Specific design techniques are developed to ensure proper operation and outperform existing regulators, as discussed in the following section.

3.3.2 Proposed Regulator Topology for Near-Threshold Computing

A simplified circuit schematic of the proposed hybrid regulator is shown in Figure 3.9. The switched-capacitor circuit and LDO operate in a parallel fashion where the source node of the power transistor within the LDO is

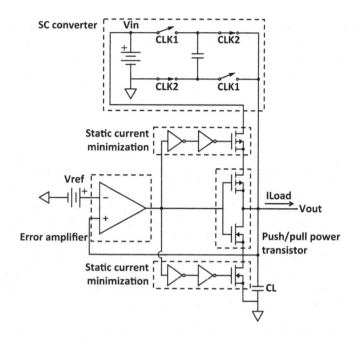

FIGURE 3.9
Proposed hybrid regulator consisting of an LDO (push/pull power transistors, error amplifier, and static current minimization) and switched-capacitor converter.

connected to the primary DC input voltage V_{in}. Thus, this topology does not suffer from the aforementioned limitations since LDO has a relatively larger input voltage. The switched-capacitor circuit provides the nominal output current while converting the input voltage from 1.15 to 0.5 V. At the nominal load current, LDO is turned off. Any variation at the output voltage is directly sensed by the error amplifier of the LDO, and output voltage is regulated with a fast transient response time. Some of the important characteristics of the proposed topology are

- No resistors are used within the LDO to minimize power loss.
- A static current minimization technique is developed to maximize power efficiency.
- Since the output voltage is directly sensed by the error amplifier, a small gain-bandwidth product is adopted, thereby preventing the output ripple from being amplified.

These characteristics are described in the following subsections.

3.3.2.1 Switched-Capacitor DC-DC Converter

A switched-capacitor converter consists of several switches and capacitors to achieve voltage conversion, as discussed in Section 3.2.2. The topology shown in Figure 3.9 achieves a conversion ratio of 2, as typically required for near-voltage computing with existing technologies where the nominal supply voltages are in the range of 0.8–1 V and threshold voltages are in the range of 0.3–0.4 V. According to [40], the overall power loss can be expressed by

$$P_{loss} = P_{C_{fly}} + P_{R_{sw}} + P_{bott-cap} + P_{gate-cap}, \tag{3.5}$$

where $P_{C_{fly}}$, $P_{R_{sw}}$, $P_{bott-cap}$, and $P_{gate-cap}$ refer, respectively, to power loss due to flying capacitor, switch resistance, parasitic capacitance of flying capacitor, and that of the switches. $P_{C_{fly}}$ and $P_{R_{sw}}$ are

$$P_{R_{sw}} \propto I_L^2 \frac{R_{on}}{W_{sw}}, \quad P_{C_{fly}} \propto I_L^2 \frac{1}{C_{fly} f_{sw}}, \tag{3.6}$$

where I_L is the load current, R_{on} is the on resistance of a single switch, W_{sw} is the width of a single switch, C_{fly} is the flying capacitance, and f_{sw} is the switching frequency. The shunt power loss due to fully integrated flying capacitor $P_{bott-cap}$ and gate capacitance of the switches $P_{gate-cap}$ are

$$P_{bott-cap} \propto V_o^2 C_{bott} f_{sw}, \quad P_{gate-cap} \propto V_{sw}^2 C_{gate} f_{sw}, \tag{3.7}$$

where C_{bott} is the sum of the parasitic capacitance from the top and bottom plates of the flying capacitor, V_{sw} is the clock voltage swing, and C_{gate} is the gate capacitance of the switches. Following these expressions, the switch size and flying capacitor are determined to maximize power efficiency [40, 41]. These parameters are listed in Table 3.1.

3.3.2.2 Resistorless LDO

The proposed LDO does not contain any resistors to maximize power efficiency, as illustrated in Figure 3.9. Instead, a PMOS push power transistor provides the additional current to the load, whereas an NMOS pull transistor reduces the output voltage. These power transistors are controlled by the output of the error amplifier. The error amplifier directly senses the output voltage and adjusts its output based on the difference between the reference voltage and output voltage. Two important design characteristics are the

TABLE 3.1
Primary Parameters of the Switched-Capacitor Converter

W_{sw}/L_{sw}	$43 \times 25 \ \mu m/50 \ nm$
C_{fly}	1.5 nF
C_L	1.5 nF
Switching frequency	482 MHz

error amplifier and the static current minimization technique, as described in the following subsections.

3.3.2.3 Optimization of the Error Amplifier

In conventional LDOs, the output frequency spectrum is determined solely by the error amplifier within the LDO. Alternatively, in the proposed regulator, the high-frequency components of the output frequency spectrum, as depicted in Figure 3.10, are dominantly determined by the switched-capacitor converter since it operates in parallel with the LDO. As listed in Table 3.1, the switching frequency is 482 MHz. According to Figure 3.10, the output voltage has a strong frequency component at this switching frequency, demonstrating the effect of the switched capacitor on the frequency spectrum. Thus, the ripple at the output voltage is primarily determined by the switched capacitor. This behavior is important since the error amplifier directly senses the output voltage in this approach. To prevent the error amplifier from amplifying output ripple, the gain-bandwidth product should be smaller than the switching frequency of the switched-capacitor circuit. Note, however, that a sufficiently small gain-bandwidth product slows down the circuit, increasing the response time. Considering this trade-off, the gain-bandwidth product is determined as approximately 350 MHz.

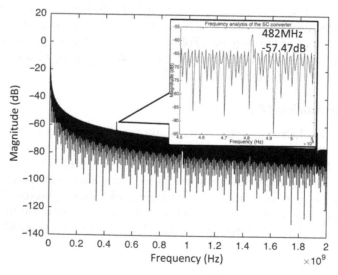

FIGURE 3.10
Frequency spectrum of the output voltage with 100 mA current.

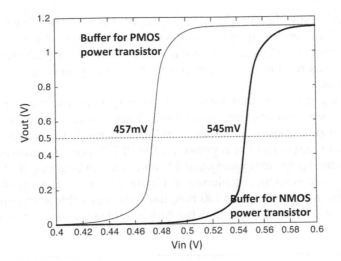

FIGURE 3.11
DC analysis of the buffers added to prevent static current.

3.3.2.4 Static Current Minimization

As opposed to traditional LDOs with single PMOS power transistor, the proposed LDO consists of both PMOS and NMOS power transistors to be able to increase and decrease the output voltage during regulation. Thus, according to the error amplifier output, both power transistors can be on, dissipating significant static current. This behavior should be prevented to maximize power efficiency. For this reason, a buffer with a different switching voltage is added before each power transistor, as illustrated in Figure 3.9. The DC voltage characteristics of these buffers are shown in Figure 3.11. As illustrated in this figure, the buffer preceding the PMOS power transistor has a much smaller switching voltage than the buffer preceding the NMOS power transistor. This difference in the switching voltage ensures that either (1) only PMOS power transistor is on or (2) only NMOS power transistor is on or both (3) power transistors are off. The difference in the switching voltages is determined to ensure that the situation when both transistors are on is avoided, thereby preventing the static current.

3.3.3 Simulation Results

The proposed novel hybrid regulator is designed using a 45 nm complementary metal-oxide semiconductor (CMOS) technology with a capacitance density of $30\,nF/mm^2$. The input voltage is 1.15 V and the output voltage is 0.5 V, which is slightly larger than the threshold voltage. The nominal load current is 100 mA, as supplied by the switched-capacitor converter. The total capacitance is 3 nF, which approximately occupies 0.1 mm², thereby achieving approximately 0.5 W/mm². As recently demonstrated by [42], regulators for portable SoCs require this power density to ensure proper operation at reasonable cost.

The output voltage and error amplifier output are plotted in Figure 3.12 when the load current varies from 65 to 130 mA. As illustrated in this figure, the output of the error amplifier is reduced as the load current increases. Thus, additional current is supplied by the PMOS power transistor. Output voltage remains approximately at 0.5 V with a maximum ripple of 22 mV.

The power efficiency is plotted in Figure 3.13 as a function of load current. At the nominal load of 100 mA, the regulator achieves approximately 85% power efficiency. Note that the power efficiency is maintained above

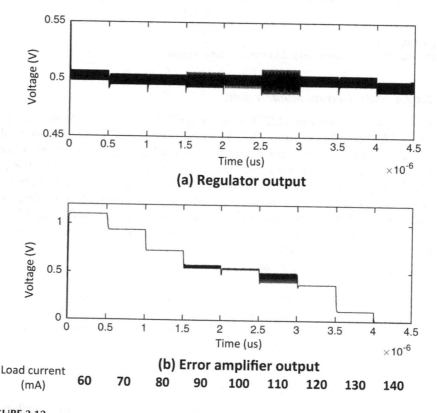

(a) Regulator output

(b) Error amplifier output

| Load current (mA) | 60 | 70 | 80 | 90 | 100 | 110 | 120 | 130 | 140 |

FIGURE 3.12

Simulation results as the load current abruptly changes from 60 to 140 mA with a step size of 10 mA: (a) output voltage of the regulator and (b) output voltage of the error amplifier.

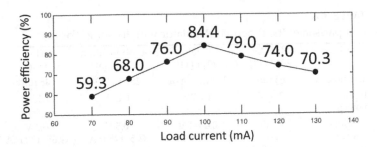

FIGURE 3.13
Power efficiency of the proposed regulator.

70% across a relatively broad range of load current, from approximately 82 to 130 mA.

Finally, the transient response of the proposed regulator is depicted in Figure 3.14. When the load current changes from 65 to 130 mA, the regulator requires approximately 18 ns to regulate the output voltage back to 0.5 V. Alternatively, when the load current changes from 130 to 65 mA, the regulator responds more quickly with a response time of 10 ns. The maximum overshoot and undershoot are less than 50 mV in both cases.

The proposed regulator is compared with several recent existing works, developed for similar applications. The comparison results are listed in Table 3.2. According to this table, at comparable current density, this work outperforms other works in both power efficiency and output ripple. Specifically, the output ripple is reduced by more than 60%, enabling a more robust near-threshold operation. A reasonable transient response time is also achieved.

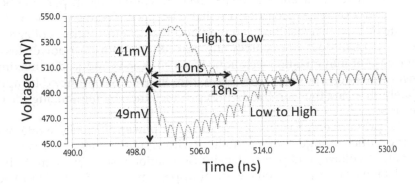

FIGURE 3.14
Transient response of the proposed regulator when the load current abruptly changes.

TABLE 3.2

Comparison of the Proposed Regulator with Existing Work

Reference	H.-P. Le 2013 [43]	R. Jain 2014 [44]	M. Abdelfattah 2015 [45]	This work
Technology	65 nm	22 nm trigate	45 nm SOI	45 nm
Input voltage	3–4 V	1.23 V	1.15 V	1.15 V
Output voltage	1 V	0.45–1 V @ 88 mA	0.5 V @ 5–125 mA	0.5 V @ 65–130 mA
Power efficiency	73%	70%@ 0.55 V 84%@ 1.1 V	74–80% @ 5–125 mA	84.4% @ 100 mA
Response time	N/A	3–5 ns	3–95 ns	<20 ns
Current density	0.19 A/mm^2	0.88 A/mm^2	1.25 A/mm^2	1 A/mm^2
Ripple voltage	N/A	60 mV	62 mV	Max: 22 mV Min: 9 mV

3.4 Summary and Conclusions

The significant opportunities and fundamental challenges related to monolithic voltage regulators have been discussed in this chapter. An overview of primary voltage regulator topologies such as linear dropout, switched-capacitor-based, and switching buck regulators has been provided. A qualitative comparison of these topologies was also considered to describe design trade-offs related to on-chip integration.

In the second part of this chapter, a novel hybrid regulator topology was described with application to near-threshold computing in portable SoCs. Contrary to existing approaches, a switched-capacitor converter and an LDO operate in a parallel fashion to convert and regulate the output voltage. The proposed LDO does not contain any resistors to minimize power loss. A static current minimization technique has also been introduced to maximize power efficiency. The error amplifier within the LDO is optimized by appropriately choosing the gain-bandwidth product, thereby minimizing the output ripple.

The primary emphasis is on maximizing power efficiency while maintaining sufficient regulation capability (with ripple voltage less than 5% of the output voltage) and power density. Simulation results in 45 nm technology demonstrate a power efficiency of approximately 85% at 100 mA load current with an input and output voltage of, respectively, 1.15 and 0.5 V. The worst-case transient response time is under 20 ns when the load current varies from 65 to 130 mA. The worst-case ripple is 22 mV while achieving a power density of 0.5 W/mm^2.

References

1. E. Salman and E. G. Friedman, *High Performance Integrated Circuit Design*, McGraw-Hill Professional, 2012.
2. G. Semeraro et al., "Energy-Efficient Processor Design Using Multiple Clock Domains with Dynamic Voltage and Frequency Scaling," *Proceedings of the IEEE International Symposium on High-Performance Computer Architecture*, pp. 29–40, Feb. 2002.
3. C. Sitik, W. Liu, B. Taskin, and E. Salman, "Design Methodology for Voltage-Scaled Clock Distribution Networks," *IEEE Transactions on Very Large Scale Integration (VLSI) Systems*, Vol. 10, No. 24, pp. 3080–3093, 2016.
4. V. Kursun and E. G. Friedman, *Multi-Voltage CMOS Circuit Design*, John Wiley & Sons, 2006.
5. E. Salman, E. G. Friedman, R. M. Secareanu, and O. L. Hartin, "Worst Case Power/Ground Noise Estimation Using an Equivalent Transition Time for Resonance," *IEEE Transactions on Circuits and Systems I: Regular Papers*, Vol. 56, No. 5, pp. 997–1004, May 2009.
6. E. Salman, E. G. Friedman, and R. M. Secareanu, "Substrate and Ground Noise Interactions in Mixed-Signal Circuits," *Proceedings of the IEEE International SoC Conference*, pp. 293–296, Sept. 2006.
7. H. Wang and E. Salman, "Closed-Form Expressions for I/O Simultaneous Switching Noise Revisited," *IEEE Transactions on Very Large Scale Integration (VLSI) Systems*, Vol. 25, pp. 769–773, Feb. 2017.
8. Z. Gan, E. Salman, and M. Stanacevic, "Figures-of-Merit to Evaluate the Significance of Switching Noise in Analog Circuits," *IEEE Transactions on Very Large Scale Integration (VLSI) Systems*, Vol. 23, pp. 2945–2956, Dec. 2015.
9. E. A. Burton et al., "FIVR—Fully Integrated Voltage Regulators on 4th Generation Intel Core SoCs," *Proceedings of IEEE Applied Power Electronics Conference and Exposition*, pp. 432–439, 2014.
10. V. Kursun, S. G. Narendra, V. K. De, and E. G. Friedman, "Efficiency Analysis of a High Frequency Buck Converter for On-Chip Integration with a Dual-V_{DD} Microprocessor," *Proceedings of the IEEE European Solid-State Circuits Conference*, pp. 743–746, Sept. 2002.
11. V. Kursun, S. G. Narendra, V. K. De, and E. G. Friedman, "Analysis of Buck Converters for On-Chip Integration with a Dual Supply Voltage Microprocessor," *IEEE Transactions on Very Large Scale Integration Systems*, Vol. 11, No. 3, pp. 514–522, 2003.
12. T. Simunic et al., "Dynamic Voltage Scaling and Power Management for Portable Systems," *Proceedings of the IEEE/ACM Design Automation Conference*, pp. 524–529, June 2001.
13. W. Kim, M. S. Gupta, G. Y. Wei, and D. M. Brooks, "Enabling On-Chip Switching Regulators for Multi-Core Processors Using Current Staggering," *Proceedings of the Workshop on Architectural Support for Gigascale Integration*, June 2007.
14. G. Patounakis, Y. W. Li, and K. Shepard, "A Fully Integrated On-Chip DC-DC Conversion and Power Management System," *IEEE Journal of Solid-State Circuits*, Vol. 39, No. 3, pp. 443–451, Mar. 2004.

15. K. N. Leung and P. K. T. Mok, "A Capacitor-Free CMOS Low-Dropout Regulator with Damping-Factor-Control Frequency," *IEEE Journal of Solid-State Circuits*, Vol. 37, No. 10, pp. 1691–1701, Oct. 2003.

16. R. K. Dokaniz and G. A. Rincon-Mora, "Cancellation of Load Regulation in Low Drop-Out Regulators," *Electronics Letters*, Vol. 38, No. 22, pp. 1300–1302, Oct. 2002.

17. C. K. Chava and J. Silva-Martinez, "A Robust Frequency Compensation Scheme for LDO Voltage Regulators," *IEEE Transactions on Circuits and Systems I: Regular Papers*, Vol. 51, No. 6, pp. 1041–1050, June 2004.

18. G. A. Rincon-Mora and P. E. Allen, "A Low-Voltage, Low Quiescent Current, Low Drop-Out Regulator," *IEEE Journal of Solid-State Circuits*, Vol. 33, No. 1, pp. 36–44, Jan. 1998.

19. P. Hazucha et al., "Area-Efficient Linear Regulator with Ultra-Fast Load Regulation," *IEEE Journal of Solid-State Circuits*, Vol. 40, No. 4, pp. 933–940, Apr. 2005.

20. D. Maksimovic and S. Dhar, "Switched-Capacitor DC-DC Converters for Low-Power On-Chip Applications," *Proceedings of the IEEE Power Electronics Specialists Conference*, pp. 54–59, June 1999.

21. A. Chandrakasan and R. W. Brodersen, *Low Power CMOS Digital Design*, Kluwer Academic Publishers, 1995.

22. D. Baba, "Benefits of a Multiphase Buck Converter," *Analog Applications*, pp. 8–13, 2012.

23. X. Zhou et al., "Investigation of Candidate VRM Topologies for Future Microprocessors," *IEEE Transactions on Power Electronics*, Vol. 15, No. 6, pp. 1172–1182, 2000.

24. J. Clarkin, "Design of a 50A Multi-Phase Converter," *Proc. Conf. Rec., HFPC*, pp. 414–420, 1999.

25. W. Kim, M. S. Gupta, G. Y. Wei, and D. Brooks, "System Level Analysis of Fast, Per-Core DVFS Using On-Chip Switching Regulators," *Proceedings of IEEE Symposium on High Performance Computer Architecture*, pp. 123–134, 2008.

26. Z. Toprak et al., "5.2 Distributed System of Digitally Controlled Microregulators Enabling Per-Core DVFS for the POWER8 TM Micro- processor," *Proceedings of IEEE International Solid-State Circuits Conference*, pp. 98–99, 2014.

27. Y. Park and E. Salman, "On-Chip Hybrid Regulator Topology for Portable SoCs with Near-Threshold Operation," *Proceedings of the IEEE International Symposium on Circuits and Systems*, pp. 786–789, 2016.

28. C. Huang and P. Mok, "An 84.7% Efficiency 100-MHz Package Bondwire-Based Fully Integrated Buck Converter with Precise DCM Operation and Enhanced Light-Load Efficiency," *IEEE Journal of Solid-State Circuits*, Vol. 48, No. 11, pp. 2595–2607, 2013.

29. Y. Ahn, H. Nam, and J. Roh, "A 50-MHz Fully Integrated Low-Swing Buck Converter Using Packaging Inductors," *IEEE Transactions on Power Electronics*, Vol. 27, No. 10, pp. 4347–4356, 2012.

30. S. A. Chickamenahalli et al., "RF Packaging and Passives: Design, Fabrication, Measurement, and Validation of Package Embedded Inductors," *IEEE Transactions on Advanced Packaging*, Vol. 28, No. 4, pp. 665–673, 2005.

31. C. Yan, Z. Gan, and E. Salman, "Package Embedded Spiral Inductor Characterization with Application to Switching Buck Converters," *Microelectronics Journal*, Vol. 66, pp. 41–47, Aug. 2017.

32. C. Yan, Z. Gan, and E. Salman, "In-Package Spiral Inductor Characterization for High Efficiency Buck Converters," *Proceedings of the IEEE International Symposium on Circuits and Systems*, pp. 2396–2399, May 2017.
33. K. Onizuka et al., "Stacked-Chip Implementation of On-Chip Buck Converter for Distributed Power Supply System in SiPs," *IEEE Journal of Solid-State Circuits*, Vol. 42, No. 11, pp. 2404–2410, 2007.
34. J. Wibben and R. Harjani, "A High-Efficiency DC-DC Converter Using 2 nH Integrated Inductors," *IEEE Journal of Solid-State Circuits*, Vol. 43, No. 4, pp. 844–854, 2008.
35. S. S. Kudva and R. Harjani, "Fully-Integrated On-Chip DC-DC Converter with a 450X Output Range," *IEEE Journal of Solid-State Circuits*, Vol. 46, No. 8, pp. 1940–1951, 2011.
36. R. G. Dreslinski, M. Wieckowski, D. Blaauw, D. Sylvester, and T. Mudge, "Near-Threshold Computing: Reclaiming Moore's Law through Energy Efficient Integrated Circuits," *Proceedings of the IEEE*, Vol. 98, No. 2, pp. 253–266, Feb. 2010.
37. B. Zhai, R. G. Dreslinski, D. Blaauw, T. Mudge, and D. Sylvester, "Energy Efficient Near-threshold Chip Multi-processing," *ACM/IEEE Int. Symp. on Low Power Electronics and Design*, pp. 32–37, Aug. 2007.
38. M. El-Nozahi, A. Amer, J. Torres, K. Entesari, and E. Sanchez-Sinencio, "High PSR Low Drop-Out Regulator with Feed-Forward Ripple Cancellation Technique," *IEEE Journal of Solid-State Circuits*, Vol. 45, No. 3, pp. 565–577, Mar. 2010.
39. K. K. G. Avalur and S. Azeemuddin, "Automotive Hybrid Voltage Regulator Design with Adaptive LDO Dropout Using Load-sense Technique," *IEEE Asia Pacific Conf. on Circuits and Syst.*, pp. 571–574, Nov. 2014.
40. H.-P. Le, S. R. Sanders, and E. Alon, "Design Techniques for Fully Integrated Switched-Capacitor DC-DC Converters," *IEEE Journal of Solid-State Circuits*, Vol. 46, No. 9, pp. 2120–2131, Sept. 2011.
41. M. D. Seeman and S. R. Sanders, "Analysis and Optimization of Switched-Capacitor DC-DC Converters," *IEEE Workshops on Computers in Power Electronics*, pp. 216–224, July 2006.
42. L. G. Salem and P. P. Mercier, "A Footprint-constrained Efficiency Roadmap for On-chip Switched-capacitor DC-DC Converters," *IEEE Int. Symp. on Circuits and Systems*, pp. 2321–2324, May 2015.
43. H. P. Le, J. Crossley, S. R. Sanders, and E. Alon, "A Sub-ns Response Fully Integrated Battery-connected Switched-capacitor Voltage Regulator Delivering $0.19W/mm^2$ at 73% efficiency," *IEEE Int. Solid-State Circuits Conf.*, pp. 372–373, Feb. 2013.
44. R. Jain, B. M. Geuskens, S. T. Kim, M. M. Khellah, J. Kulkarni, J. W. Tschanz, and V. De, "A 0.45-1 V Fully-Integrated Distributed Switched Capacitor DC-DC Converter with High Density MIM Capacitor in 22 nm Tri-Gate CMOS," *IEEE Journal of Solid-State Circuits*, Vol. 49, No. 4, pp. 917–927, Apr. 2014.
45. M. Abdelfattah, B. Dupaix, S. Naqvi, and W. Khalil, "A Fully-integrated Switched Capacitor Voltage Regulator for Near-threshold Applications," *IEEE Int. Symp. on Circuits and Systems*, pp. 201–204, May 2015.

4

Low-Power Biosensor Design Techniques Based on Information Theoretic Principles

Nicole McFarlane

CONTENTS

4.1 Introduction

Mixed-signal complementary metal-oxide semiconductor (CMOS) technology has become a popular research area for integrated biosensing applications. However, while modern CMOS processes, through the fulfillment of Moore's law, realize decreasing minimum sizing, this is accompanied by a lessening power supply. Further, the inherent physical noise still remains the same. This trend leads to poor signal-to-noise ratios and dynamic range performance being significant challenges to sensitive and accurate low-power biosensing. The application of information theory to circuits has been introduced to model various topologies. These topologies include chopper stabilized amplifiers, active pixel sensors, and single-photon avalanche diodes [1–6]. By using circuit design methodologies based on information theory, it is possible to create mixed-signal systems that can operate at lower power, while efficiently transmitting information in the presence of high intrinsic physical and environmental noise. The methodology, and its implications, in this chapter was previously presented as part of [7].

4.2 Noise and Information Rates in Amplifiers

Noise, from a variety of sources, distorts signals of interest, setting a lower bound on the minimum detectable signal (Figure 4.1). Additionally, biosignals tend to be extremely small and hover around the levels of intrinsic system noise. The intrinsic transistor noise sources include thermal noise, a white band noise source with a constant value spectral density, and flicker noise, which varies inversely with the frequency, gate current noise, and shot noise. The flicker noise and thermal noise sources are typically dominant, particularly at the lower frequencies where biosignals are found. The output noise spectral density for the thermal and flicker noise current sources is

$$S_{I_d} = \gamma 4KT g_m + \frac{K_f I_d^{A_f}}{f^{E_f} C_{ox} L_{eff}^2} \tag{4.1}$$

where the K_f, A_f, and E_f parameters are dependent on the fabrication process [8–12]. γ depends on the region of operation. In strong inversion it has

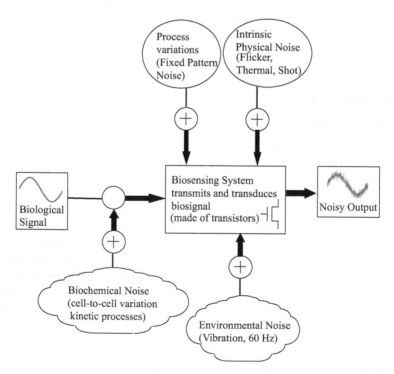

FIGURE 4.1
Overview of possible noise sources in biosensing microsystems.

a value of 2/3, while in weak inversion it has a value of 1/2. The other intrinsic noise sources, gate leakage current nosie and shot noise, are typically nondominant sources of noise for metal-oxide-semiconductor field-effect transistors (MOSFETs). p-MOSFETs typically have lower noise than n-MOSFETs in a given process. The output current noise is transformed to voltage noise through the equivalent output impedance [13–15],

$$S_{out_v} = S_{I_d} Z_{out}^2. \tag{4.2}$$

For any mixed-signal, multistage system, including amplifiers, the voltage noise at the input is more important to consider than the output noise. The noise at the input node of any stage is determined from the gain of the output signal with respect to the input signal. For multiple stages, the noise is divided by the gain for each stage until the input node is reached. Process parasitics, including the gate-to-source and gate-to-drain capacitances (C_{gd} and C_{gs}), may need to be considered to accurately account for frequency effects. Since each device is assumed to have an independent noise generator, the total noise density is simply the sum of all the noise sources. Thus, the total input noise at the input of the amplifier is determined by developing the effect of each noise source on the output divided by the frequency-dependent gain of the amplifier.

A method for extracting and measuring noise characteristics of transistors is shown in Figure 4.2 [16,17]. Amplifier noise is measured at the output directly using the spectrum analyzer and a voltage-buffering circuit to prevent the analyzer from loading the amplifier. The voltage buffer and amplifier should be supplied using a noiseless power source such as a battery with voltage regulation. The measured experimental noise for an n-channel transistor in a standard CMOS process is shown in Figure 4.3 [7].

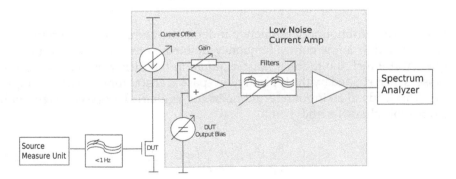

FIGURE 4.2
Setup for noise parameter K_f and A_f, extraction. DUT is the device under test.

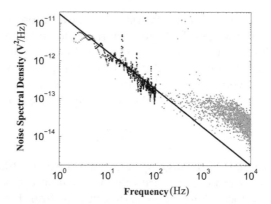

FIGURE 4.3
Flicker (1/f) noise profile of an NMOS transistor with $K_f = 10^{-26}$ [7].

4.3 First-Order Model for the Information Rate of an Amplifier

For differential amplifiers, the noise referred to the input terminals is dominated by the initial stage, and specifically by the input differential pair. The voltage noise source at the input of the amplifier can be represented as

$$S_{Vg} = \frac{4KT\gamma}{g_m} + \frac{K_f}{WLC_{ox}fg_m^2} \tag{4.3}$$

More generically, following [18], this has the form

$$S_{Vg} = S_o \left(1 + \frac{f_k}{f}\right) \tag{4.4}$$

where S_0 is the amount of frequency-independent white noise from thermal sources, and f_k is the corner frequency defined as the point where the frequency at which the thermal noise is equal to the flicker noise. The output noise is then shaped by the typically low-pass, transfer function of the amplifier. This leads to an input referred noise for a generic low-pass system with single-pole characteristics [19]:

$$S_{n_{in}} = \frac{S_0}{A_0^2} \left(1 + \frac{f_k}{f}\right) \left(1 + \left(\frac{f}{f_c}\right)^2\right) \tag{4.5}$$

f_k is the corner frequency as previously defined, S_0 is the noise from thermal sources, A_0 is the midband gain of the system, and f_c is the bandwidth of the system. These parameters are functions of the physical makeup, including the fabrication process, of the transistors. This includes the noise parameters for flicker noise, K_f, A_f, and E_f); the transconductance, g_m; the input differential pair transistor aspect ratio, W/L; the output resistance, r_o; and the bias current from the tail current source, I.

In biological applications, for example, in electrophysiological experiments, the input biological signal presented to the amplifier is extremely weak. The assumption is made that the noise sources are Gaussian in nature. Following classic information theory, the amplifier can be appropriately modeled as a Gaussian communication channel. The solution to maximizing the information efficiency of this channel is well known as the waterfilling technique, where the input signal is spectrally diffused over frequency locations or bins where the noise is at a minimum [20,21]. This may be mathematically expressed as

$$C = \int_{f_1}^{f_2} \log_2 \left(\frac{v}{S_{n_{in}}(f)} \right) df \tag{4.6}$$

v is the power spectral density of the input signal power plus the spectral density of the flicker and thermal noise signals for the system's bandwidth. $\Delta f = f_2 - f_1$ is a fixed number, and the input signal power is defined as

$$P = \int_{f_1}^{f_2} (v - S_{n_{in}}(f)) \, df$$
$$v = S_{n_{in}}(f_1) = S_{n_{in}}(f_2) \tag{4.7}$$

Assuming an ideal, single-pole, low-pass transfer function for the filter, the information rate is

$$v = S_{n_{in}}(f_1) \text{ and } f_2 = f_c = f_{3dB}$$

$$I = \frac{1}{\ln 2} \left[2(f_c - f_1) + f_c \ln \frac{\left(1 + \frac{f_k}{f_1}\right)\left(1 + \frac{f_1^2}{f_c^2}\right)}{\left(1 + \frac{f_k}{f_c}\right) 2} \right.$$

$$\left. + f_k \ln \frac{f_1 + f_k}{f_c + f_k} - 2f_c \left(\frac{\pi}{4} - \tan^{-1} \frac{f_1^2}{f_c^2} \right) \right] \tag{4.8}$$

The bandwidth and the signal power are given by

$$P_{sig} = \frac{S_0}{A_0^2} \left[\frac{2}{3} \frac{f_2^3 - f_1^3}{f_c^2} + \frac{f_k}{2f_c^2} \left(f_2^2 - f_1^2 \right) - f_k \ln \frac{f_1}{f_2} \right] \tag{4.9}$$

$$0 = \left(1 + \frac{f_k}{f_2}\right)\left(1 + \frac{f_2^2}{f_c^2}\right) - \left(1 + \frac{f_k}{f_1}\right)\left(1 + \frac{f_1^2}{f_c^2}\right) \tag{4.10}$$

As defined previously, in waterfilling, the power of the input signal is first allocated to frequency bins where the thermal + flicker noise power is at a minimum, before being allocated to frequencies where these noise sources are higher. In this manner, as the signal power increases, so does the information rate. Assuming the flicker-thermal corner frequency lies within the bandwidth, the first two quantities rely only on the frequency where the noise corner occurs and system bandwidth. Increasing the bandwidth or the frequency of the noise corner results in an information rate increase. Typically, $f_1 \ll f_c$ and $f_2 \approx f_c$; therefore, as the bandwidth, f_c, increases, $arctan\,(f_2/f_c)$ remains almost at a fixed value, while $arctan\,(f_1/f_c)$ linearly increases. For a fixed noise corner frequency, f_k, larger-bandwidth systems will have greater information rates. As the bandwidth increases, $f_c \gg f_{1,2}$, the last two terms cancel. With lowered corner frequencies, $f_c < f_{1,2}$, the sum approaches $2\,(f_2 - f_1)$.

Most systems will have multiple poles and zeroes; for example, most two-stage amplifiers will have at least two dominant poles and a zero. Other poles and zeroes are generally high enough to be neglected for the amplifier performance. The information rate can be generalized for a system with poles and zeroes as

$$
C \;=\; \frac{1}{\ln 2}\left[2(f_2 - f_1) + f_k \ln \frac{f_1 + f_k}{f_2 + f_k} \right.
$$
$$
- \sum_{i=1}^{n} 2f_{pi}\left(\tan^{-1}\frac{f_2}{f_{pi}} - \tan^{-1}\frac{f_1}{f_{pi}} \right)
$$
$$
\left. + \sum_{j=1}^{m} 2f_{zj}\left(\tan^{-1}\frac{f_2}{f_{zj}} - \tan^{-1}\frac{f_1}{f_{zj}} \right) \right] \tag{4.11}
$$

where $f_{1,2}$ are determined by the shape of the spectrum for the power density of the noise sources and the quantity of available signal power. Clearly, higher signal powers will require higher bandwidth (subject to the same of the spectral density function). Each additional pole decreases the information rate, while zeroes increase the information rate. A decrease in information capacity will be dominated by the dominant pole, while any increase will be dominated by the first zero. From this result, signal power should be place at frequencies above the dominant pole, where any amplification is typically minimal. This, thus requires careful design strategies to optimize signal power, noise power, and power supply using information rate and typical application design constraints.

The input referred noise is lowered with decreasing bias currents due to reduced bandwidth and increased midband gain. The dominant pole location is a function of the bias current; thus, lowered bias current also influences the frequency location of the noise minimum. Thus, the water-filling frequency allocation of signal power may vary significantly with the bias current. If the bandwidth is about the same or smaller than the frequency of the thermal-flicker noise intersection, the frequency location of the minimum value of noise may increase.

4.4 Metrics for Trade-Offs in Power and Noise

There are many other metrics for characterization of the interplay noise of a system and the available power specifications. Specifically, the noise efficiency factor (NEF) is a popular metric for characterizing noise and power of a system. It compares the noise of the given system with an ideal bipolar transistor. The ideal bipolar is assumed to have no base resistance and only thermal noise sources [22]. Given some frequency bandwidth, Δf; bipolar collector current, I_c; and total current draw from the power supplies of the amplifier, I_{tot}, the NEF is given by

$$\text{NEF} = \sqrt{\Delta f \frac{\pi}{2} \frac{4kTV_T}{I_c}} \sqrt{\frac{2I_{tot}}{\pi 4kTV_T \Delta f}} \qquad (4.12)$$

where V_T is the thermal voltage, 0.025 V. A less noise- and power-efficient amplifier will have a greater NEF value. Improved system noise performance is implied by having a lowered NEF.

Another metric known as the bit energy, BE, measures the energy cost of using the system with optimized amplifier performance [23]. This is measured by the power consumed by the amplifier, P_{amp}, and maximum information rate or information capacity, I, such that

$$BE = \frac{P_{amp}}{I} \qquad (4.13)$$

Lower bit energy implies a more efficient amplifier in terms of cost to utilize the system and performance. That is, the cost of maximizing amplifier performance will be minimized.

The total power in an amplifier is a function of the power supplies, V_{dd} and V_{ss}, as well as the total current for the amplifier and any biasing and start-up circuitry,

$$P_{amp} = I_{tot} \left(V_{dd} - V_{ss} \right) \qquad (4.14)$$

The power increases as the current increases. However, the dominant pole also increases with the current. This leads to less efficient operation, as the maximum information transfer rate decreases and the cost of using the system increases. The NEF is related to the bit energy as they both include the total current and noise. However, the bit energy takes the available power resources explicitly into account in its formulation, making it a more accurate measure.

4.5 Trade-Offs in a Simple Amplifier Design Example

A simple operational transconductance amplifier (OTA) is shown in Figure 4.4. The major design parameters and constraints for an amplifier using the EKV model are the bandwidth, midband gain, and intersection of the flicker and thermal spectral density plots. In the EKV model, these are all formulated in terms of the inversion coefficient, IC, as [24]

$$A_o = \frac{\kappa}{U_T} \frac{1 - e^{-\sqrt{IC}}}{\sqrt{IC}} V_{A_{2,4}}$$

$$S_o = \frac{4kT\gamma}{\frac{\kappa I_D}{U_T} \frac{1 - e^{-\sqrt{IC}}}{\sqrt{IC}}}$$

$$f_k = \frac{K_f}{WLC_{ox}4kT \frac{\kappa}{U_T} \frac{1 - e^{-\sqrt{IC}}}{\sqrt{IC}}}$$

$$f_{3dB} = \frac{IC I_o W/L}{2\pi V_{A_{2,4}} C_{out}}$$

$$V_{A_{2,4}} = \frac{V_{A_2} V_{A_4}}{V_{A_4} + V_{A_2}}$$

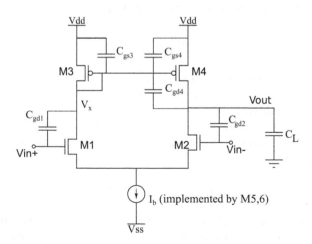

FIGURE 4.4
Simple differential OTA explicitly showing the parasitic capacitances between the terminals of the transistor [7].

where the inversion coefficient is defined as the ratio of the drain current to the reverse saturation current, I_d/I_s. W/L is the transistor aspect ratio, and V_A is the early voltage (reciprocal of λ, the transistor channel length modulation parameter). $U_T = V_T$ is the thermal voltage (0.025 V), C_{ox} is the capacitance of the gate oxide, I_o is the characteristic current, and κ is the slope in subthreshold. The simplified model of γ uses a constant of $1/2$ for transistors operating above threshold, and a value of $2/3$ for transistors operating in strong inversion. Over weak, moderate, and strong inversion regions, a more complete model is [24]

$$\gamma = \frac{1}{1+IC}\left(\frac{1}{2}+\frac{2}{3}IC\right) \tag{4.15}$$

The design parameters to be determined are the bias tail current, typically implemented as a current mirror source with M_5 and M_6, and transistor aspect ratios. The midband gain at lowered frequencies is traditionally written as

$$A_0 = g_{m1}\left(r_{o2} \parallel r_{o4}\right) = \sqrt{\frac{\mu C_{ox}W/L}{I_{bias}}\frac{1}{\lambda_2+\lambda_4}} \tag{4.16}$$

in saturation. The MOSFET square law equations are assumed and λ is the inverse of the early voltage. The dividing line that separates the weak, moderate, and strong inversion lies at inversion coefficient values of 0.1, 1, and 10.

The minimum inputs occur due to the tail current transistor entering into triode (assuming strong inversion):

$$V_{I_{min}} \geq \sqrt{\frac{I_B}{\beta_1}}+V_{th1}+\sqrt{\frac{2I_B}{\beta_5}}+V_{ss} \tag{4.17}$$

The maximum input signal occurs when M_4 enters triode and eventually turns off, so that

$$V_{G1} = V_{DD}-\sqrt{\frac{I_b}{\mu C_{ox}W/L}}-V_{th3}+V_{th1} \tag{4.18}$$

The common mode gain is a function of the tail current output resistance,

$$A_c = \frac{1}{2g_{m4}r_{o5}} = \frac{1}{\frac{4\kappa 1-e^{-\sqrt{IC}}}{U_T\sqrt{IC}}} \tag{4.19}$$

And the common mode rejection ratio (CMRR) is given by $\left|\frac{A_v}{A_c}\right|$. The slew rate increases with increasing bias current and lowered load capacitances. The load capacitance consists of parasitic capacitances and any explicit capacitances at the output node. Increased current increases both the thermal and flicker noise levels. These noise levels decrease with transistor area. Thus, increasing inversion coefficient increases the noise as well as the thermal and flicker noise corner frequency, system bandwidth, and cost of using the system (bit energy). Gain decreases with decreasing inversion coefficient. However, the system information rate shows a complex response to design parameters [1].

For the amplifier, ignoring process variations means $g_{m1} = g_{m2}$ and $g_{m3} = g_{m4}$. The input referred noise, assuming no frequency effect, is [25]

$$v_{eq}^2 = \frac{i_{n1}^2}{g_{m1}^2} + \frac{i_{n2}^2}{g_{m1}^2} + \left(\frac{g_{m3}}{g_{m1}}\right)^2 \left(\frac{i_{n3}^2}{g_{m3}^2} + \frac{i_{n4}^2}{g_{m3}^2}\right) \tag{4.20}$$

Figure 4.4 shows the parasitic capacitances that affect the transfer function. The capacitances of the NMOS and PMOS pair, from the gate to the source and from the gate to the drain, are assumed to be equal. The sources of differential pair, M1 and M2 are considered to be at virtual ground, and since the bias current adds equally to both sides, it is not considered to add to the system noise. The system transfer function from input to output nodes, and from the gate of the current mirror active load to the input, is

$$H(f) = $$

$$\frac{-1/2\left(sC_{gd1} - g_{m1}\right)\left(sC_{gd1} + sC_1 + 1/R_1 + g_{m4}\right)}{\left(sC_{out} + 1/R_{out}\right)\left(sC_{gd1} + 1/R_1 + sC_1 + sC_{gd4}\right) - sC_{gd4}\left(sC_{gd4} - g_{m4}\right)}$$

$$\frac{V_x}{V_{in}} = $$

$$\frac{1/2\left(sC_{gd1} - g_{m1}\right)\left(sC_{gd4} + sC_out + 1/R_{out}\right)}{sC_{gd4}\left(sC_{gd4} - g_{m4}\right) - \left(sC_{out} + 1/R_{out}\right)\left(sC_{gd1} + 1/R_1 + sC_1 + sC_{gd4}\right)}$$

the poles and zeroes of the amplifier, assuming C_{gd4} is small, are

$$z_1 = \frac{g_{m1}}{C_{gd2}}$$

$$z_2 = \frac{2g_{m4}}{C_1 + C_{gd1}}$$

$$p_1 = \frac{1}{R_o\left(C_L + C_{gd2}\right)}$$

$$p_2 = \frac{g_{m4}}{C_1 + C_{gd1}} \tag{4.21}$$

TABLE 4.1

Variation of Design Specification with Inversion
Coefficient (IC) and Length (L)

Specification	Name	IC ↑	L ↑
A_o	Low frequency gain	↓	↑
f_k	Noise corner frequency	↑	↓
S_o	White noise value	↑	↑
A_c	Common mode gain	↑	↓
BE	Bit energy	↑	↓
I	Information rate	↑	↓

where:

$$C_1 = C_{gs3} + C_{gs4}$$

$$C_{out} = C_L + C_{gd2} + C_{gd4}$$

$$R_1 = r_{o1}||r_{o3}||1/g_{m3}$$

$$R_{out} = r_{o2}||r_{o4} \tag{4.22}$$

Given that $i = g_m v_{gs}$, the noise voltage is reflected back to the gate of the transistor as i_n^2/g_m^2 at low frequencies. At midfrequencies, the gate-to-drain capacitance may not necessarily be ignored, and the noise voltage at the gate is $i_n^2/(g_m + sC_{gd})^2$.

The width of a transistor may be taken out of the design space by considering instead that the bias current, inversion level, and transistor length are known [26, 27]. Bias current and length of the transistor both change linearly with the inversion level. Both aspect ratio, W/L, and area, $W \times L$, impact the system and should be explored first, after which the effects of length and bias current are explored [26].

Increased inversion coefficient, which is moving from weak to moderate to strong inversion, affects the information rate and energy cost of using the system (bit energy). The stronger the inversion level, the better the information rate, implying that an above-threshold operation will give a cleaner transfer of the input signal to the measured output signal. The cost of using the amplifier is lowered at the lower inversion levels. Since the input differential pair should contribute the most noise, a generalization can be made of the design specifications as functions of the transistor length and region of operation (Table 4.1).

4.6 Information Rate of Amplifier

Capacitive feedback with an OTA is a popular architecture for weak biosignal acquisition [3, 28, 29]. The design has been widely implemented in various CMOS processes. A wide-range OTA is used as the OTA for the system shown in Figure 4.5. For a typical design in a 500 nm process, the input differential pair transistors have W = 35 μm, L = 2.1 μm, C2 = 200 fF, and C1 = 20 pF. For a 130 nm process, the input differential pair transistors have W = 24 μm, L = 1.2 μm, C2 = 98.3 fF, and C1 = 10 pF. The power supply goes from 5 to 1.8 V between the two processes. The pseudoresistors have a large resistance value. The noise spectrum referred to the input terminals of the OTA is a function of the input capacitance (C_{in}) and the output noise (S_{OTA}), as is [28],

$$S_{amp} = \left(\frac{C_1 + C_2 + C_{in}}{C_1} \right)^2 S_{OTA} \qquad (4.23)$$

Figure 4.6 shows the measured noise at the output, frequency-dependent magnitude of the transfer function, spectral density of the noise referred to the input terminals, information rate, and energy cost of using the amplifier [7]. Due to a wider minimal noise spectral density bandwidth, the larger process shows a higher information rate than the shorter channel process. There is a slight decrease in voltage between the processes; however, it is unable to compensate for the noise-driven information rate measure. With the fundamental noise sources actually increasing with decreasing process lengths, this result shows that the expected improvement in the energy costs (due to lower

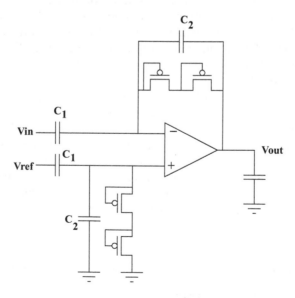

FIGURE 4.5
Frequently used capacitive feedback amplifier topology used for biosignal acquisition [28].

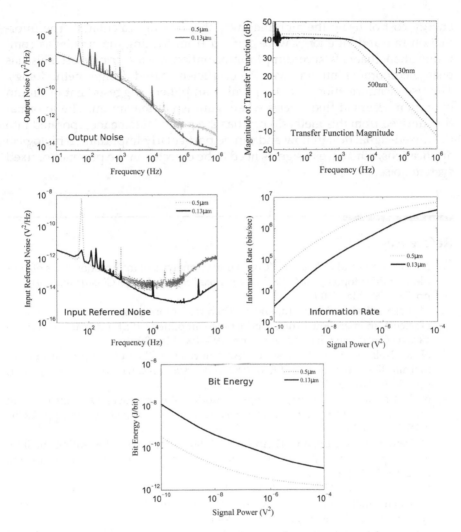

FIGURE 4.6
Comparison of larger and submicron process output voltage noise, input referred noise, transfer function, information rates, and bit energy [7].

voltage supplies) of submicron and deep submicron processes does not materialize.

4.7 Conclusion

The input referred thermal and flicker noise spectral density for an amplifier used for weak biological signals, along with the information rate and energy cost of using the amplifier, was theoretically calculated. The information rate and bit energy were derived from treating the circuits as communication channels according to information theory principles. This has been incorporated into an inversion coefficient–based design methodology. The use of information rate to optimize amplifiers suggests that optimum operation occurs at frequencies where gain is not maximum. The information derived from this methodology has the potential to be incorporated into the mixed-signal design flow and can be particularly important in biological applications, where weak signals need to be detected in the presence of fixed system noise.

References

1. N. McFarlane and P. Abshire, "Comparative analysis of information rates of simple amplifier topologies," *IEEE International Symposium of Circuits and Systems*, pp. 785–788, May 2011.
2. M. Loganathan, S. Malhotra, and P. Abshire, "Information capacity and power efficiency in operational transconductance amplifier," *IEEE International Symposium on Circuits and Systems*, vol. 1, pp. 193–196, May 2004.
3. N. M. Nelson and P. A. Abshire, "An information theoretic approach to optimal amplifier operation," *IEEE Midwest Symposium on Circuits and Systems*, vol. 1, pp. 13–16, Aug. 2005.
4. N. Nelson and P. Abshire, "Chopper modulation improves OTA information transmission," *IEEE International Symposium on Circuits and Systems*, pp. 2275–2278, May 2007.
5. D. Sander, N. Nelson, and P. Abshire, "Integration time optimization for integrating photosensors," *IEEE International Symposium on Circuits and Systems*, pp. 2354–2357, June 2008.
6. J. Gu, M. H. U. Habib, and N. McFarlane, "Perimeter-gated single-photon avalanche diodes: An information theoretic assessment," *IEEE Photonics Technology Letters*, vol. 28, no. 6, pp. 701–704, Mar. 2016.
7. "Information power efficiency tradeoffs in mixed signal CMOS circuits," PhD dissertation, University of Maryland, College Park, 2010.
8. M. J. Kirton and M. J. Uren, "Noise in solid-state microstructures: A new perspective on individual defects, interface states and low-frequency (1/f) noise," *Advances in Physics*, vol. 38, no. 4, pp. 367–468, 1989.
9. I. Bloom and Y. Nemirovsky, "1/f noise reduction of metal-oxide-semiconductor transistors by cycling from inversion to accumulation," *Applied Physics Letters*, vol. 58, pp. 1664–1666, 1991.
10. E. A. M. Klumperink, S. L. J. Gierink, A. P. Van der Wel, and B. Nauta, "Reduction MOSFET 1/f noise and power consumption by switched biasing," *IEEE Journal of Solid-State Circuits*, vol. 35, no. 7, pp. 994–1001, 2000.
11. Y. Isobe, K. Hara, D. Navarro, Y. Takeda, T. Ezaki, and M. Miura-Mattausch, "Shot noise modelling in metal oxide semiconductor field effect transistors under sub threshold condition," *IEICE Transactions on Electronics*, vol. E90-C, no. 4, pp. 885–894, Apr. 2007.

12. L. Callegaro, "Unified derivation of Johnson and shot noise expressions," *American Journal of Physics*, vol. 74, pp. 438–440, May 2006.
13. S.-C. Liu, J. Kramer, G. Indiveri, T. Delbruck, and R. Douglas, *Analog VLSI: Circuits and Principles*, Cambridge, MA: MIT Press, 2002.
14. C. Jakobson, I. Bloom, and Y. Nemirovsky, "1/f noise in CMOS transistors for analog applications from subthreshold to saturation," *Solid-State Electronics*, vol. 42, pp. 1807–1817, 1988.
15. Y. Nemirovsky, I. Brouk, and C. G. Jakobson, "1/f noise in CMOS transistors for analog applications," *IEEE Transactions on Electron Devices*, vol. 48, no. 5, pp. 212–218, May 2001.
16. R. Tinti, F. Sischka, and C. Morton, "Proposed system solution for 1/f noise parameter extraction," http://literature.cdn.keysight.com/litweb/pdf/5989-9087EN.pdf.
17. A. Blaum, O. Pilloud, G. Scalea, J. Victory, and F. Sischka, "A new robust on-wafer 1/f noise measurement and characterization system," *International Conference on Microelectronic Test Structures*, pp. 125–130, 2001.
18. C. C. Enz and G. C. Temes, "Circuit techniques for reducing the effects of op-amp imperfections: autozeroing, correlated double sampling, and chopper stabilization," *Proceedings of the IEEE*, vol. 84, no. 11, pp. 1584–1614, Nov. 1996.
19. C. C. Enz, E. A. Vittoz, and F. Krummenacher, "A CMOS chopper amplifier," *IEEE Journal of Solid-State Circuits*, vol. 22, no. 3, pp. 335–342, June 1987.
20. T. M. Cover and J. A. Thomas, *Elements of Information Theory*, New York: John Wiley & Sons, 1991.
21. P. A. Abshire, "Implicit energy cost of feedback in noisy channels," *IEEE Conference on Decision and Control*, vol. 3, pp. 3217–3222, Dec. 2002.
22. M. S. J. Steyaert, W. M. C. Sansen, and C. Zhongyuan, "A micropower low-noise monolithic instrumentation amplifier for medical purposes," *IEEE Journal Solid-State Circuits*, vol. 22, no. 6, pp. 1163–1168, Dec. 1987.
23. A. G. Andreou, An information theoretic framework for comparing the bit energy of signal representations at the circuit level, in *Low-Voltage/Low Power Integrated Circuits and Systems*, Piscataway, NJ: IEEE Press, 1999, pp. 519–540.
24. C. C. Enz, F. Krummenacher, and E. A. Vittoz, "An analytical MOS transistor model valid in all regions of operation and dedicated to low-voltage and low-current applications," *Analog Integrated Circuits and Signal Processing*, vol. 8, no. 1, pp. 83–114, 1995.
25. P. R. Gray, P. J. Hurst, S. H. Lewis, and R. G. Meyer, *Analysis and Design of Analog Integrated Circuits*, New York: John Wiley & Sons, 2001.
26. D. Binkley, B. Blalock, and J. Rochelle, "Optimizing drain current, inversion level, and channel length in analog CMOS design," *Analog Integrated Circuits and Signal Processing*, vol. 47, pp. 137–163, 2006.
27. D. Binkley, C. Hopper, S. Tucker, B. Moss, J. Rochelle, and D. Foty, "A CAD methodology for optimizing transistor current and sizing in analog CMOS design," *IEEE Transaction on Computer-Aided Design*, vol. 22, pp. 225–237, 2003.
28. R. R. Harrison and C. Charles, "A low-power low-noise CMOS amplifier for neural recording applications," *IEEE Journal of Solid-State Circuits*, vol. 38, no. 6, pp. 958–965, June 2003.
29. S. B. Prakash, N. M. Nelson, A. M. Haas, V. Jeng, P. Abshire, M. Urdaneta, and E. Smela, "Biolabs-on-a-chip: Monitoring cells using CMOS biosensors," *IEEE/NLM Life Science Systems and Applications Workshop*, pp. 1–2, July 2006.

5

A Cost-Effective TAF-DPS Syntonization Scheme of Improving Clock Frequency Accuracy and Long-Term Frequency Stability for Universal Applications

Liming Xiu, Pao-Lung Chen, and Yong Han

CONTENTS

5.1 Introduction

Expression $S(t) = A(t) \cdot \sin\{2\pi f_0 \cdot [1 + D(t)] \cdot t + \varphi(t)\}$ describes a sinusoidal signal S(t) outputted from any frequency source where f_0 is the nominal center frequency. The weak time dependence of amplitude A(t) and phase offset $\varphi(t)$ makes them primarily short-term fluctuations (jitter/phase noise). The drift D(t) produces a slow change on center frequency (such slow change is called "frequency wander"). Frequency *accuracy* is the term used to describe the closeness of a frequency source's center frequency f_0 to its specified (desirable) value. Fractional frequency is defined as $(f_0 - f)/f_0$, where f is the frequency at any given moment. The *precision* on frequency measurement depends on the time τ taken to make the measurement. The measurement result is usually displayed graphically as fractional frequency fluctuation $\sigma_y(\tau)$, which is commonly known as Allan variance. Frequency *stability* describes the frequency source's capability of maintaining its frequency within a specified range around the f_0. In industry, there are four types of frequency sources with different quality in term of stratum level: 1, 2, 3, and 4. In hardware implementation, frequency standards include cesium, rubidium, and crystal with frequency stability ranging from $1.0 \cdot 10^{-11}$ to $1.6 \cdot 10^{-8}$ to $4.6 \cdot 10^{-6}$ to $3.2 \cdot 10^{-5}$ (corresponding to the four stratum levels) [1–3].

Frequency links *time* through the clock. Time is produced by a clock signal and is used to coordinate events inside an electronic system. In a network structure, timing is always passed from clocks of a given stratum performance level to clocks of a lower or equal stratum performance level [4–9]. Clock signal can be directly generated from a frequency source, or indirectly through a phase-locked loop (PLL). Frequency source is found in military, metrology, industrial, consumer, communication network, automotive, power grid, banking, and scientific research electronic systems. In all those systems, frequency stability influences the system performance from multiple directions: the slip rate in digital communication (such as STN, NTP, and PTP) [6, 8], the precision in navigation systems (such as GPS) [9, 10], the time accuracy in bookkeeping (such as in power grids, banking, and stock trading) [11, 12], and the improvement in military warfare (e.g., improved spectrum utilization, higher jamming resistance, fast signal acquisition, longer life, and smaller size, weight, or cost) [13], among others. In addition, with the adoption of atomic standards and the complexity of today's sophisticated applications, modern systems also need to consider the impact of relativity on the accuracy of the clock [14, 15].

A typical clock system includes a frequency source and a counting circuit (for counting, setting, and synchronization). There are two concepts when clocks are used to coordinate events among systems: synchronization and syntonization. Clocks are *synchronized* when they agree in *time*. Clocks are *syntonized* when their oscillators have the same frequency. No clock can ever keep perfect time since all oscillators exhibit random and systematic errors.

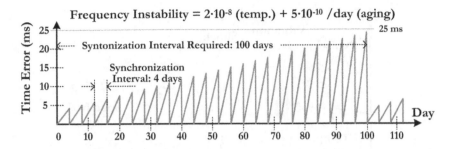

FIGURE 5.1
Frequency instability leads to time error.

Clock error can be expressed in Equation (5.1), where T(t) is the time difference between two clocks at time t after synchronization, T_0 is synchronization error at t = 0, R(t) is the frequency difference between the two clocks' oscillators, and $\varepsilon(t)$ is the error due to random fluctuations in the system.

$$T(t) = T_0 + \int_0^t R(t)dt + \varepsilon(t) \tag{5.1}$$

When compared with the ideal source, the frequency value of a real source always deviates from the specified value. This imperfection, which is caused by the combined effect of manufacture error, temperature variation, aging, and loading shift, shows its effect as frequency instability. It can negatively impact system performance. An example is given below to illustrate the point. Assume that a system requires a 25 ms time accuracy for its operation. Also assume that its frequency source has a frequency offset of $2 \cdot 10^{-8}$ due to temperature and, in addition, an aging rate of $5 \cdot 10^{-10}$/day. Initially, the clock is assumed to have zero frequency and time error. The clock is resynchronized every 4 days. Figure 5.1 shows the growing trend of time error. As shown, the time error is caused almost entirely by the frequency offset of $2 \cdot 10^{-8}$ during the early phase. Each day, however, aging adds $5 \cdot 10^{-10}$ to the frequency error. At 40th days, the frequency error due to aging equals that due to temperature. After 40 days, aging becomes the dominant cause and the time error increases more and more during each 4-day resynchronization interval. Eventually, after about 100 days, the time error at the end of the 4-day-period reaches the 25 ms limit. At that point, either the resynchronization interval must be shortened or the clock must be resyntonized (oscillator's frequency must be readjusted to the correct value). If the aging rate, however, is changed from $5 \cdot 10^{-10}$/day to $5 \cdot 10^{-11}$/day, the time for syntonization could be relaxed from 100 days to 3 years (under the same condition of resynchronization every 4 days) [16].

Equation (5.1) relates frequency error to time error. Figure 5.2 shows the major causes of frequency instability: temperature step, aging, vibration, manufacture error, shock, oscillator on/off switching, pressure, humidity,

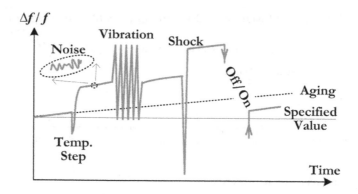

FIGURE 5.2
Major types of frequency instabilities.

etc. [16]. In a high-performance system, a temperature-compensated crystal oscillator (TCXO) or even an oven-controlled crystal oscillator (OCXO), is used. Those high-quality sources usually have much better frequency stability than low-end crystals. The drawback, however, is the extreme high cost. Out of the causes depicted in Figure 5.2, the frequency instabilities due to manufacture error, temperature, and aging can potentially be compensated. Commercial examples for compensating temperature-induced instability are available. Two examples are given in [17, 18], where temperature sensors are used to report temperature reading and built-in fractional-N or integer-N PLL is used to counteract the corresponding frequency deviation. Those compensations, however, target particular applications and the circuits are custom designed case by case. It is not a general approach for dealing with the frequency error problem.

The study on time/clock/frequency is a subject of multiple disciplines, including physics, communication, computer science, and very large-scale integration (VLSI) circuit design. A generic method for alleviating the frequency error problem is our challenge. We need a solution aiming for universal applications. This task has to be fulfilled regarding the fact that the structure of the frequency source cannot be altered once it is manufactured and mounted in the final system. For this reason, it is desirable to have a means for compensating the frequency error/draft in a normal operation environment so that the user can carry out the frequency compensation task whenever necessary. This can both lower system cost in the design phase and prolong the product's lifetime in field usage. Considering the number of electronic systems used in our modern life, its economic impact cannot be overstated.

Time-average-frequency direct period synthesis (TAF-DPS) is an emerging frequency synthesis technique [19, 20]. Its distinguishing features are small frequency granularity (also termed arbitrary frequency generation) and fast frequency switching (also termed instantaneous frequency switching). Experimental evidence is available to support the claim that its

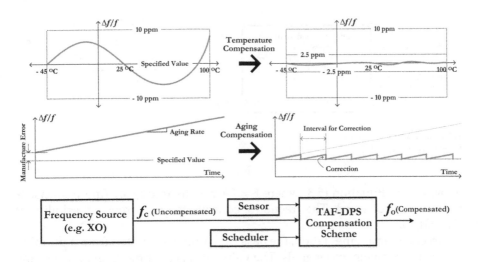

FIGURE 5.3
TAF-DPS method of counteracting frequency error.

frequency granularity can reach the level of a few parts per billion (ppb) [21]. In this work, TAF-DPS is used to develop a method for correcting frequency error on the field without physically altering the frequency source. Figure 5.3 shows our plan for accommodating manufacture error, temperature-induced frequency error, and aging-related frequency deviation.

In this chapter, Section 5.2 is devoted to the discussion of circuit architecture for improving frequency. It also includes an exemplary case of implementation on a field-programmable gate array (FPGA). Section 5.3 provides experimental data for supporting the proposed architecture. Section 5.4 presents the TAF-DPS syntonization scheme for improving long-term frequency stability. Section 5.5 concludes the chapter.

5.2 Architecture of Frequency Fine Tuning

5.2.1 Brief Review of TAF-DPS

$$T_A = I \cdot \Delta, \quad T_B = (I+1) \cdot \Delta \tag{5.2}$$

$$T_{TAF} = 1/f_{TAF} = (1-r) \cdot T_A + r \cdot T_B = (I+r) \cdot \Delta = F \cdot \Delta \tag{5.3}$$

Figure 5.4 shows the principal idea of TAF-DPS. It is based on the TAF concept [19]. From a base time unit Δ, the synthesizer first creates two (or more) types of cycles T_A and T_B. Their length in time are given by Equation (5.2) where I is an integer. When synthesizing a particular frequency f_{TAF}, it uses T_A and T_B in an interleaved fashion. The output frequency/period is

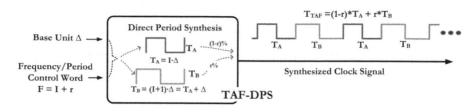

FIGURE 5.4
Principle of TAF-DPS.

expressed in Equation (5.3) where $F = I + r$ is the frequency (more precisely, period) control word. The fraction r controls the occurrence possibility of T_A and T_B.

In circuit implementation, the base time unit Δ is created from a plurality of phase-evenly-spaced signals. The value of Δ is the time span between any two adjacent ones of such a group of signals. This group of signals can be the outputs of a multistage voltage-controlled output that is locked to a reference through a PLL. Or, they can simply be generated from a divider chain. The DPS circuit is an edge selector and combiner. Experimental evidence of ~2 ppb frequency granularity is presented in [21].

5.2.2 Architecture of Improving Frequency Accuracy

Figure 5.5 depicts our architecture of using TAF-DPS for compensating the frequency error associated with a frequency source. The frequency source has a specified value of f_{c_target}, while its actual output is f_c. In most cases, $f_c \neq f_{c\_target}$ due to various reasons, such as manufacture error, component aging, or temperature deviating from normal value. The difference between specified and actual values could be as large as a few hundreds parts per millions (ppm). Thanks to its excellent frequency tunability, TAF-DPS is used here to compensate this frequency error. Its purpose is to make $f_o = f_{c\_target}$ and, hopefully, make this true under all disturbances from the surrounding environment.

The output of the frequency source is used to drive a base time Δ generator whose output is fed into the TAF-DPS synthesizer. The Δ generator

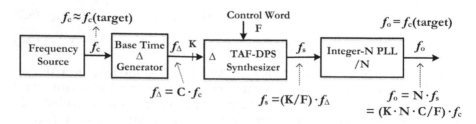

FIGURE 5.5
Architecture.

can be created from a PLL or DLL. If a PLL of divide ratio M is used, f_Δ = M·f_c, while $f_\Delta = f_c$ if DLL is used. Further, a divider chain of dividing ratio K (e.g., a Johnson counter of K/2 stages, as in the example in [21]) can also be used. It results in a plurality of K phase-evenly-spaced signals at frequency $f_\Delta = f_c/K$. In general, $f_\Delta = C·f_c$, where C is a constant. From Equation (5.3) where f_{TAF} is the f_s in this case, it is derived that $f_s = (K/F)·f_\Delta$. Therefore, the final output f_o is related to the frequency source's output f_c as $f_o = (K·N·C/F)·f_c$.

$$\Delta = T_c, \quad f_s = f_c/F, \quad f_o = (N/F) \cdot f_c \qquad (5.4)$$

In the case of a divider chain, C = 1/K and $\Delta = T_\Delta/K = 1/(f_\Delta·K) = 1/f_c = T_c$. Those relationships are listed in Equation (5.4). As seen, it is possible to make f_o (= $(N/F)·f_c$) = f_{c_target} since F can take a fractional value. The following numerical example is helpful to clear the picture. Assume the frequency source is a 100 MHz crystal. At one moment, the measured value is f_c = 99.999723 MHz, which is about 2.77 ppm (~277 Hz) off target. If the PLL's N is set as 16 and the TAF-DPS frequency control word F is set as F = 15.99995568, according to Equation (5.4), the final output f_o is calculated as f_o = (16/15.99995568)*99.999723 = 99.999999 MHz. Hence, the frequency error has been reduced from 2.77 ppm to virtually none.

5.2.3 Definition and Analysis

Definitions

f_{c_target}	The specified frequency value of the frequency source (the target value)
f_c	The actual measured value of the frequency source, $f_{c\_target} = (1+x)·f_c$, where x is the error
f_o	The compensated frequency value, the final output
F	The TAF-DPS frequency control word, F = I + r where I is an integer and —r— $\leq$ 0.5
N	An integer, the PLL divide ratio

From Equation (5.4) and using the above definitions, Equation (5.5) can be derived:

$$f_o/f_c = f_o/[f_{c\_target}/(1+x)] = (N/F) = N/[I \cdot (1+r/I)] \qquad (5.5)$$
$$\to 1+x = 1/(1+r/I) = 1 - r/I + (r/I)^2 - (r/I)^3 + \dots$$
$$x \approx -r/I \qquad (5.6)$$

The goal of our compensation process is to make $f_o = f_{c\_target}$. Further, we usually set I = N for convenience. Thus, to cancel the frequency error, we use Equation (5.6) to calculate the value of r to the first order (the value of r/I is usually much smaller than 1%). The negative sign is because TAF-DPS's output frequency is inversely proportional to the control word (refer to Equation (5.3)). For the example given above, x = 2.77 ppm (positive sign indicates that the output f_c is slower than the target value → need to speed up). For I = 16, the calculated fraction is $r = -x \cdot I = -2.77 \cdot 10^{-6} \cdot 16 = -0.00004432$. This leads to F = I + r = 15.99995568, as used in the previous example. In the operation, the value of I is first chosen by the user and r is subsequently determined from the required amount of correction x.

5.2.4 Discussion on Time-Average-Frequency Signal

In this architecture, it is worth pointing out that the signal at the TAF-DPS output (f_s in Figure 5.5) uses TAF, while outputs at all other points are conventional frequencies. The integer-N PLL converts the TAF signal into a conventional frequency signal for the final output (refer to Figure 5.5). Interestingly, the two controllable variables, I and r, lead to two different styles of TAF implementations: type L of large {I, r} values and type S of small {I, r} values. It results in different modulation strengths, which influences the operation of the following PLL.

As can be seen in Equations (5.2) and (5.3), I controls the similarity (or dissimilarity) of T_A and T_B, while r is in charge of their occurrence possibility. For type L, it uses T_B more often (r is larger). Further, large I makes T_B more look like T_A since $T_B = (1+1/I) \cdot T_A$. On the other hand, for type S, smaller I drives T_B away from T_A, and it also forces T_B be used less frequently. In this regard, it could be said that type L waveform is more *regular* (it resembles the conventional clock waveform more) and thus less modulation is created. To describe this phenomenon more precisely, a parameter called *irregularity ρ* is defined to measure the "distance" between the waveform of TAF and that of conventional frequency. ρ is calculated using Equation (5.7):

$$\rho = \frac{2/\pi - \|V_{fa}\|}{2/\pi} \cdot 100 \tag{5.7}$$

where $\|V_{fa}\|$ is the amplitude of the TAF tone while $2/\pi$ is the amplitude of the corresponding tone of conventional frequency. Details on ρ are available in Section 5.8 of [20]. Thus, the smaller the ρ is, the more regular is the TAF waveform, and the weaker is the modulation. Table 5.1 presents the characteristics.

5.2.5 Case of Implementation on FPGA

A low-cost Atlys™ FPGA system is used to verify the architect presented in Figure 5.5. The Atlys circuit board is a complete, ready-to-use digital circuit development system based on a Xilinx Spartan-6 LX45 FPGA, speed grade 3 [22]. A 100 MHz crystal from this board serves as the frequency source to be compensated. The divider chain is implemented as a Johnson counter, created from the configurable FPGA elements. In addition, there are multiple

TABLE 5.1

Characteristics of Time-Average-Frequency Implementation

	Type L: larger {I, r}	Type S: smaller {I, r}
T_A and T_B similarity	↑	↓
T_A and T_B occurrence	more even	less even
Waveform regularity	more regular, ρ ↓	less regular, ρ ↓
Modulation strength	weak	strong
TAF-DPS frequency f_s	↓	↑
Required PLL ratio N	↑	↓

PLLs available from this FPGA system as standard components. Therefore, one of them is selected to function as the integer-N PLL.

The TAF-DPS is also implemented from the configurable FPGA elements. The circuit for implementing this TAF-DPS is chosen as Flying-Adder frequency synthesis architecture [23–26]. Its circuit block diagram is shown in Figure 5.6. Interested readers are referred to [24] for circuit details. In this

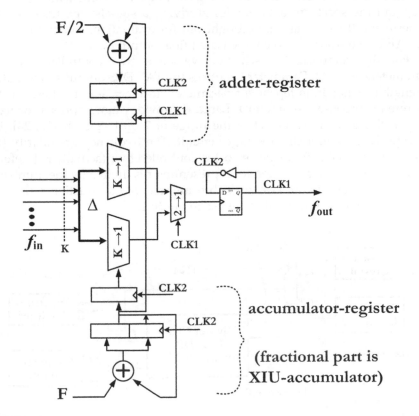

FIGURE 5.6

Block diagram of Flying-Adder synthesizer.

particular case, the implementation approach is HDL coding → simulation → synthesis and map to FPGA. The VHDL code for this Flying-Adder circuit is available in Appendix 4.A of [20].

5.3 Experiment Data

5.3.1 Experiment Setup

Figure 5.7 illustrates the test setup for verifying our frequency compensation architecture. The frequency source is either a crystal or a signal generated from a signal generator. Crystal is used to demonstrate that the frequency accuracy of a source can be improved by our circuit. The signal generator, due to its capability of varying frequency as "frequency wander," is needed to study certain scenarios (such as frequency drift). The test equipment used are the frequency counter for measuring the frequency value with high accuracy and the spectrum analyzer for studying a signals' spectrum. In our experiment, three signals are brought out for evaluation: source frequency f_c, TAF-DPS output f_s, and compensated final output f_o.

For the compensation scheme, we have two controllable design parameters: K and I. Parameter I controls the TAF-DPS output's modulation strength (r is not an independent parameter since $r = -x \cdot I$, where x is the required compensation amount). Large K provides more options for us to play with I since I can be set in the range of $2 \leq I \leq 2 \cdot K$ [20, 24]. But a larger K leads to a slightly larger circuit. The frequency granularity (the minimal correctable frequency error) is controlled by r (actually r/I; refer to Equation (5.6)). The resolution for representing r is limited by the number of bits reserved for r. In our system, 32 bits are reserved for fraction r. Table 5.2 lists several configurations that we use in our test.

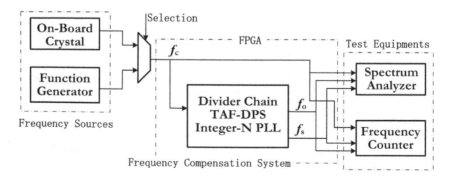

FIGURE 5.7
Test setup.

5.3.2 Improvement on Frequency Accuracy

To verify the effectiveness of our scheme for improving a source's frequency accuracy, multiple tests using the configurations listed in Table 5.2 have been carried out in the lab. A Keysight (Agilent) 53220A frequency counter is used to measure the frequencies of f_c, f_s, and f_o. Figure 5.8 is the screen capture that corresponds to case 4 in Table 5.2 (it is also the numerical example discussed in Section 5.2.2). As seen in the picture, the original crystal output is measured as 99.9997232665373 MHz (CH2). The final output after compensation is measured as 100.000000579855 MHz (CH1). Therefore, the accuracy is improved from 2.77 ppm to 5.79 ppb. This is an improvement of 478 times. As displayed, the ratio of CH1/CH2 is 1.00000277, which confirms the relation of $f_{c\_target} = (1+x) \cdot f_c$, where x = 2.77 ppm was used in the calculation for setting r.

TABLE 5.2

Test Configurations (32 bits reserved for fraction r)

Configuration Case	K	I (= N)	r[a]	f_s MHz[b] (target)	Granularity at 24th bit: $2^{-24}/I$	ρ[c]
2	16	4	$1.108 \cdot 10^{-5}$	25	14.9 ppb, 1.5 Hz	9.97
3	16	8	$2.216 \cdot 10^{-5}$	12.5	7.45 ppb, 0.75 Hz	2.55
4	16	16	$4.432 \cdot 10^{-5}$	6.25	3.73 ppb, 0.37 Hz	0.64
5	16	30	$8.31 \cdot 10^{-5}$	3.3333	1.98 ppb, 0.2 Hz	0.18
6	32	30	$8.31 \cdot 10^{-5}$	3.3333	1.98 ppb, 0.2 Hz	0.18
7	32	50	$1.385 \cdot 10^{-4}$	2	1.19 ppb, 0.11 Hz	0.07
8	32	62	$1.718 \cdot 10^{-4}$	1.6129	0.96 ppb, 0.1 Hz	0.04

[a] r is set for compensating 2.77 ppm error.

[b] For the case of f_{c_target} = 100 MHz crystal.

[c] Calculated from {I, r} (refer to [20] if interested).

FIGURE 5.8
Accuracy improved to 5.79 ppb, measured by Keysight 53220A.

To check the validity of our method further, a PICOTEST frequency counter is used to redo the same test. On the left of Figure 5.9, the crystal output f_c is measured as 99.9997637416 MHz. This is about 2.36 ppm

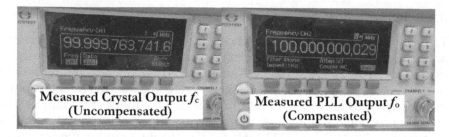

FIGURE 5.9
Accuracy improved to 0.29 ppb, measured using PICOTEST® U6200A.

slower than the target ($x = 2.362589581 \cdot 10^{-6}$). From Equation (5.6), r is calculated and the TAF-DPS control word is set as F = 15.999962198566704 ($r \approx -0.000037801433296$). This control word is fed into the TAF-DPS, and the resulting final output is measured as $f_o = 100.000000029$ MHz, as displayed on the right-hand side of Figure 5.9. Hence, the accuracy is improved from 2.36 to 0.29 ppb, an improvement of almost five orders of magnitude.

5.3.3 Evaluation of Frequency Compensation Range

$$f_{c\_target}/f_c = 1 + x \approx 1 - r/I + (r/I)^2$$

$$\rightarrow f_o/f_c \approx 1 - r/I + (r/I)^2 \text{ since we want } f_o = f_{c\_target} \quad (5.8)$$

From the definition in Section 5.2.3 and Equation (5.6), we have Equation (5.8). It is seen that the range for compensation is controlled by r/I. Lab tests have been carried out to evaluate the compensation range. The results are then compared with the predication from Equation (5.8) and presented in Table 5.3. The first column is the value of r. Several values, ranging from 2^{-24} to 2^{-4}, are tested. The case of 2^{-24} demonstrates the capability for smaller frequency correction (3.7 ppb), while 2^{-4} is for large correction (0.39%). The second and third columns are the calculation and measured results, respectively. It is worth mentioning that we only test to 2^{-4} since it is already 0.39% (3900 ppm), which is more than enough for most practice scenarios. From a circuit perspective, however, TAF-DPS has a much larger frequency tuning range.

Figure 5.10 shows the case of $r = -2^{-24}$ and $F = I - r = 16 + 2^{-24}$, which results in $f_o/f_c = 1.0000000037253$. The measured value, as displayed, is 1.0000000074945. In addition, Table 5.3 gives the jitter measurement result. The jitters of the source (the crystal f_c) are listed on the left, and the jitters of the final output (frequency-compensated f_o) are on the right. It is seen that the jitters of the two signals f_c and f_o are at the same order of magnitude. In other words, our compensation scheme does not degrade the clock signal's quality in a noticeable way.

TABLE 5.3

Measured f_o/f_c and Jitter at Various Settings ($I = 16, f_{c\_target} = 100$ MHz)

r	$f_o/f_c=1+r/I+(r/I)^2$	f_o/f_c(meas.)	f_c jitter (ps) rms	pk-pk	f_o jitter (ps) rms	pk-pk
2^{-24}	0.9999999962747[a]	0.9999999999421	29	210	25	205[c]
	1.0000000037253[b]	1.0000000074945	66	493	26	259
2^{-20}	0.9999999403954	0.9999999777260	27	283	36	371
	1.0000000596047	1.0000000695398	20	166	19	142
2^{-16}	0.9999990463266	0.9999990440000	24	176	22	156
	1.0000009536752	1.0000019060000	26	205	25	190
2^{-8}	0.9997559189796	0.9997559189052	21	142	19	210
	1.0002442002296	1.0004885196137	20	151	21	351
2^{-4}	0.9961090087891	0.9961089480000	24	303	29	244
	1.0039215087891	1.0078740156180	21	156	23	205

[a] The case of $F = I + r = 16 + 2^{-24}$.

[b] The case of $F = I - r = 16 - 2^{-24}$.

[c] Clock quality: 205 ps/10 ns = 2.1% (i.e., pk-pk jitter is 2.1% of its period).

FIGURE 5.10

Case of $f_o/f_c = 1 + 2^{-24}/16 = 1.0000000037253$ (calculation).

5.3.4 Frequency Spectrum

Figures 5.11 through 5.13 are the measured spectrum plots of TAF-DPS output f_s for cases 3, 5, and 8 listed in Table 5.2. In all those plots, the amount of frequency correction is $x = r/I = 2.77$ ppm and the span is set at 2 KHz so that details can be studied. For those cases, I is chosen as 8, 30, and 62, respectively. Due to the difference in the values of $\{I, r\}$, the resulting irregularities are $\rho = 2.55, 0.18$, and 0.04. In each figure, the value on the left is obtained by setting of r = 0 (conventional frequency), while the one on the right is set at the target value (TAF). The plots confirm our predication that a smaller ρ value leads to weaker modulation, which is easier for following PLL to lock to the main tone.

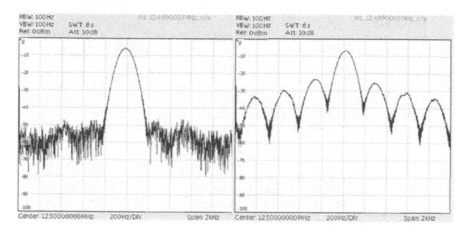

FIGURE 5.11
Measured spectrum of case 3: F = 8 (left), and F = 7.99997784, I = 8, f_{mod} = 277 Hz, ρ = 2.55 (right).

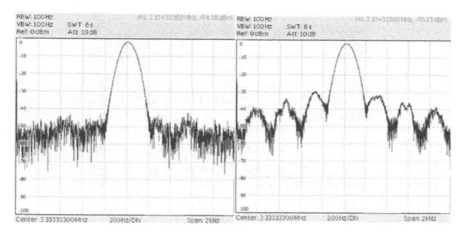

FIGURE 5.12
Measured spectrum of case 5: F = 30 (left), and F = 29.9999169, I = 30, f_{mod} = 277 Hz, ρ = 0.18 (right).

On the right-hand side of the figures, the TAF-induced spurious tones f_{mod} are all seen as spaced at ~277 Hz. In fact [20], f_{mod} can be calculated as $f_{mod} = r \cdot f_s = r \cdot (K/F) \cdot f_\Delta = x \cdot I \cdot (K/F) \cdot f_c/K \approx x \cdot f_c \approx 2.77$ ppm $*$ 100 MHz = 277 Hz. This is the case for all three plots. The bandwidth of the PLL is about 300 KHz. Therefore, the f_{mod} of 277 Hz can be traced and followed by the PLL (and hence correct frequency can be expected). In other words, the PLL output frequency is locked at the main tone of the TAF signal. Figure 5.14 shows the spectrum of f_o (the PLL output) of cases 3 and 5, and the span is set at 200 KHz. Qualitatively speaking, the two clock signals have

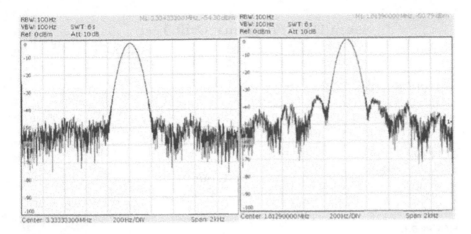

FIGURE 5.13
Measured spectrum of case 8: F = 62 (left), and F = 61.9998283, I = 62, f_{mod} = 277 Hz, ρ = 0.04 (right).

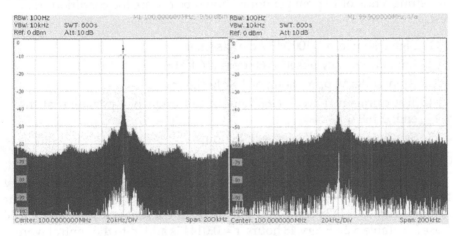

FIGURE 5.14
Spectrum of PLL output f_o: case 3 (left) and case 5 (right).

roughly the same noise level. Case 3 has slightly more modulation, which can be understood from the comparison of Figures 5.11 and 5.12.

5.3.5 Compensation for Frequency Drift (Aging)

To mimic the effect of frequency draft due to various reasons (such as component aging), a signal generator is designed and is used to generate a pulse signal whose frequency varies slowly with time in certain patterns. To simplify the test, we assume that aging-induced frequency deviation behaves linearly with a fixed rate (refer to Figure 5.3). Further, we speed up the test

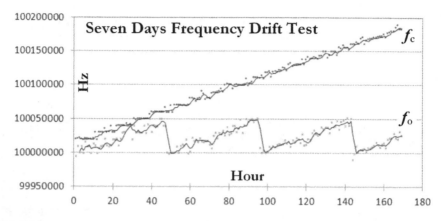

FIGURE 5.15
Frequency drift test, measuring f_c and f_o for 1 week.

by setting a rate of 10 ppm per hour. Shown below are the conditions for our test.

1. Signal source: a $\sim$100 MHz clock signal
2. Initial frequency offset: +160 ppm (16 KHz)
3. Aging rate: 10 ppm/hour, or 1 KHz/hour
4. Test duration: 1 week (168 hours); full frequency deviation at the end: 0.168%
5. Syntonization interval: every 48 hours
6. Measurements taken: every 1 hour

During the test, the signal generator's output is adjusted up 1 KHz every hour. As a result, its output frequency drifts away from 100 MHz in a linear fashion. The TAF-DPS compensation circuit is chosen as K = 16 and I = 30 (case 5 in Table 5.2). Every 48 hours, r = 0.0144 is added to the control word F ($r = 30 * 480$ ppm). This will slow down the f_o frequency at a rate of 480 ppm every 48 hours to counteract the f_c drift. Additionally, 0.0048 is added to the F in the beginning to cancel the initial frequency offset. Figure 5.15 is the measured f_c and f_o for a duration of 1 week (168 hours). Measurements are carried out once per hour. As seen, the uncompensated f_c drifts away from 100 MHz continuously. Every 48 hours, a syntonization process kicks in. After each syntonization, the compensated output f_o is brought back to 100 MHz. Between the syntonization intervals, however, f_o drifts freely.

5.3.6 Required Resource and Integration of Our Scheme in a System

Table 5.4 lists the resources used for our TAF-DPS frequency compensation scheme, for the two cases of K = 16 and 32 (the integer-N PLL is not included

TABLE 5.4

Hardware Resource Used

Number of	K = 16	K = 32
F/F	89	101
BUF	4	5
INV	6	7
LUTs	79	92
MUXs	6	5
DCM_SP	1	1
PLL_ADV	1	1

here; refer to Figure 5.5). From this table, it can be said that the required resource is very limited.

To improve electronics' timekeeping quality, a solution is to incorporate our frequency compensation scheme into the vast majority of functional chips. The mergence has to be implemented in an economic way without a high price tag. Figure 5.16 depicts the idea of integrating the scheme into a generic system-on-a-chip (SoC), which can be built in either application-specific integrated circuit (ASIC) or FPGA fashion. As demonstrated, our scheme can be completely implemented in digital style plus a standard PLL. It uses only a few hundred gates (Table 5.4). Moreover, integer-N PLL is usually available from an ASIC library as foundry IP, or from a FPGA system as a standard component. Therefore, it is a relatively easy task to include our scheme in a SoC. It is believed that the required resource for our scheme is negligible in a large SoC environment. In most cases, it can be considered "almost free." As shown in Figure 5.16, the compensated frequency f_o will be used as the reference source for all other timing circuits (function PLLs).

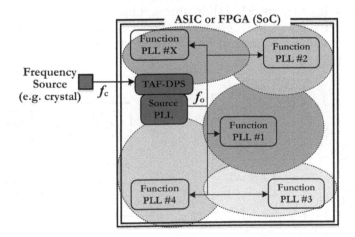

FIGURE 5.16

Incorporating frequency compensation scheme in SoC design.

5.4 TAF-DPS Syntonization for Improving Long-Term Frequency Stability

It has been demonstrated in Section 5.3 that frequency accuracy can be improved using our scheme. Further, frequency wander can be compensated. This opens up the possibility for us to improve a frequency source's long-term frequency stability by using syntonization. This section is devoted to this discussion. It is, however, worth mentioning that short-term frequency stability (jitter/phase noise) is primarily determined by device physics, and therefore it cannot be dealt with by our scheme.

5.4.1 Improve Long-Term Stability by Syntonization

Modern electronics are all expected to connect to a network as nodes, such as a computer network, telecommunications network, electrical power network, or navigation network. Nodes in networks require regular time synchronization so that a "common view of time" can be established among the nodes. The key in network synchronization is *frequency equality*. Due to various uncertainties in a network, phase lock can only be achieved on a local basis. Only frequency control is feasible on a global basis. In networks, time and/or frequency transfer is carried out in a hierarchical structure using a master-slave clock link. A master clock is usually in the high stratum level (e.g., Primary Reference Clock of stratum 1 in AT&T's network [7]). A stratum 1 clock, such as the Cesium Standards, has an excellent long-term frequency stability of 10^{-13}.

Passive atomic standards are created from the structure depicted on the left-hand side of Figure 5.17 [1]. In atomic timing devices, as seen from the structure in the figure, the high-quality clock output is achieved by a combination of crystal's good short-term stability and atomic's excellent long-term

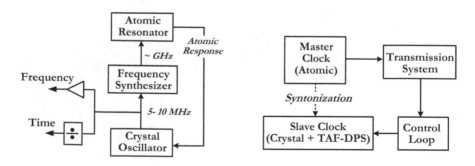

FIGURE 5.17
The short-term stability of crystal is combined with the long-term stability of atomic standards: atomic standards (left) and salve clock (right).

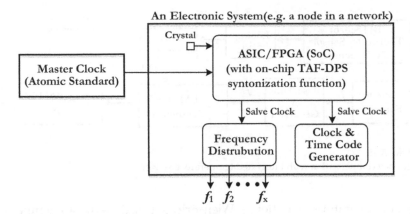

FIGURE 5.18
Salve clock of improved long-term stability used in a network.

stability. As a result, this high-quality output can be used as the master clock in networks. In our TAF-DPS syntonization scheme, a similar principle of combining crystal's good short-term stability with the master clock's excellent long-term stability is adopted. On the right-hand side of Figure 5.17, this idea is illustrated. As shown, the improvement is achieved through syntonization: by syntonizating the slave clock (crystal plus TAF-DPS) to the high-quality master clock. There are multiple ways that this syntonization task can be carried out. An interesting approach based on adaptive neural-fuzzy inference is presented in [27].

Figure 5.18 shows the block diagram in implementation. For each node in a network, this design utilizes the good short-term stability of crystal (which locates locally) and the high long-term stability of the master clock (which is available globally). As a result, the local time and frequency quality is improved for each node. Further, the improved frequency/time information can be transferred to other places in the lower hierarchy levels of the network.

5.4.2 Dealing with Temperature-Induced Frequency Variation

A crystal oscillator is the dominant frequency source used in commercial systems, primarily due to its low cost and good short-term frequency stability. Its output frequency, however, is temperature dependent, which can vary more than 100 ppm in low and high temperature extremes. A solution to this problem is temperature compensation by employing a temperature sensor in the system. A BJT-based sensor is known to achieve the best combination of accuracy and energy efficiency. It can be integrated in a system to provide temperature readings and subsequently to direct the frequency correction. An example is presented in [18]. A thermistor-based sensor can achieve

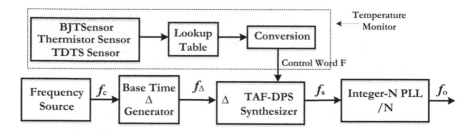

FIGURE 5.19
Frequency compensation with integrated temperature sensor.

higher temperature resolution. A Wien-Bridge-based thermistor temperature sensor is presented in [28]. A time-domain temperature sensor (TDTS) could be implemented in digital fashion and is thus low cost, low power, and more robust against process variation [29–32]. It has also been used to compensate crystal oscillators [33]. These temperature-sensing techniques can be readily used with our TAF-DPS scheme to counteract temperature-induced frequency variation, as depicted in Figure 5.19.

5.4.3 Integration with Self-Temperature Sensing

The TAF-DPS scheme can easily work with external temperature sensors, as discussed in Section 5.4.2. Moreover, it is an excellent tool to be integrated with crystal's self-temperature feature for the next-generation microcomputer-controlled crystal oscillator (MCXO).

$$f_1(T) = f_1(T_0) + a_1\Delta T + b_1\Delta T^2 + c_1\Delta T^3 + ... \tag{5.9}$$
$$f_3(T) = f_3(T_0) + a_3\Delta T + b_3\Delta T^2 + c_3\Delta T^3 + ...$$

$$f_b(T) = f_b(T_0) + (3a_1 - a_3)\Delta T + (3b_1 - b_3)\Delta T^2 + (3c_1 - c_3)\Delta T^3 + ... \tag{5.10}$$

$$f_b(T) \approx f_b(T_0) + (3a_1 - a_3)\Delta T \tag{5.11}$$

In [34], crystal's higher-order temperature coefficients have been studied. The behavior of SC-cut crystal's fundamental and third tones is illustrated in the left of Figure 5.20. When f versus T characteristics are described using polynomials, it is found that the change between the fundamental and third (and other higher) overtones is due almost entirely to the first-order temperature coefficient. Using polynomials, the temperature dependences of fundamental f_1 and third overtone f_3 can be described by Equation (5.9), where a, b, and c are constants, T_0 is a reference temperature, and $\Delta T = T - T_0$. A beat frequency can be defined as $f_b = 3f_1 - f_3$ and subsequently be derived

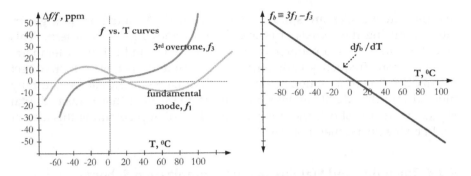

FIGURE 5.20
SC-cut crystal's fundamental and third overtone f versus T curves (left) and the beat frequency f_b versus T curve (right).

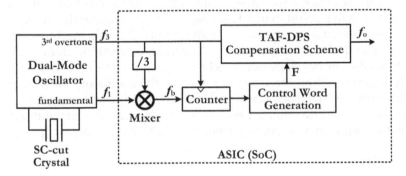

FIGURE 5.21
Architecture of self-temperature sensing combined with TAF-DPS frequency compensation scheme.

in Equation (5.10). For a temperature range of practical use, the first term is much greater than the higher-order terms [34]. Therefore, f_b varies almost linearly with temperature, as described in Equation (5.11) and illustrated on the right-hand side of Figure 5.20.

f_b is a monotonic and nearly linear function of temperature. It is a measure of the resonator's temperature exactly where the resonator is vibrating. It can eliminate the need for a thermometer. Because the SC-cut is thermal transient compensated, the thermal transient effects are also eliminated, as are the effects of temperature gradients between the thermometer and the resonator. For this reason, the beat frequency can function as an excellent temperature sensor. Hence, instead of f versus T curve, a f versus f_b curve can be created and used later for temperature-related frequency compensation.

MCXO is the type of high-stability frequency source that utilizes this self-temperature sensing [35–37]. In operation, the beat frequency f_b curve can be generated by a dual-mode oscillator where the fundamental and third overtone frequencies are simultaneously excited. A signal at f_b is generated by mixing the signals of f_3 and f_1 and removing the sum term. This signal is counted by a counter whose output is the indication of temperature (and thus frequency variation). This information is then used to direct the frequency compensation process. The TAF-DPS scheme can be naturally merged into the MCXO architecture due to the reason that it takes digital input. Figure 5.21 shows the circuit architecture of implementing this idea. It could be the next-generation MCXO.

5.4.4 Tunability and Stability of Our Syntonization Scheme

Tunability and stability are two opposite sides of a frequency source. Making an oscillator tunable over a wide frequency range degrades its stability since making an oscillator susceptible to intentional tuning also makes it susceptible to factors that result in unintentional tuning. The wider the tuning range, the more difficult it is to maintain a high stability. Our scheme, however, does not need to deal with this dilemma. As can be understood from Figure 5.5, the tunability of our approach is achieved in the TAF-DPS, which is physically separated from the frequency source (the oscillator). As a frequency synthesizer, TAF-DPS has a very wide frequency adjustable range. This wide range has no negative impact on our scheme's frequency stability since its frequency adjustment is accomplished through digital processing.

5.5 Conclusion

Considering the growing number of electronic devices supporting modern society, and further considering the fact that most of those devices are networked and need to be time synchronized to a common view of time, frequency stability will be one of the key challenges in future electronic system design. This problem, however, is nontrivial. It is a multidiscipline challenge that requires effort from computer scientists, network architects, VLSI circuit designers, and device physicists, among others. The syntonization scheme presented in this work is a first step in this direction. It is therefore expected to exert a profound influence in the decades to come. The uniqueness of our scheme is its effectiveness in frequency tuning and flexibility in implementation. The cost of incorporating it into a system is extremely low, while the improvement on time and frequency quality is very noticeable. It is hence desirable for the scheme be included in every future SoC as a standard function, to be used for universal applications. It is the authors' belief that this is a decent solution for an important problem of large economic impact.

References

1. L. L. Lewis, "An Introduction to Frequency Standards," *Proc. of the IEEE*, vol. 79, pp. 927–935, 1991.
2. P. Forman, "Atomichron: The Atomic Clock from Concept to Commercial Product," *Proc. of the IEEE*, vol. 73, no. 7, pp. 1181–1204, July 1985.
3. F. L. Walls and D. W. Allan, "Measurement of Frequency Stability," *Proc. of the IEEE*, vol. 73, no. 7, pp. 1181–1204, July 1985.
4. D. B. Sullivan, "Time Generation and Distribution," *Proc. of the IEEE*, vol. 79, pp. 906–914, 1991.
5. B. W. Petley, "Time and Frequency Fundamental Metrology," *Proc. of the IEEE*, vol. 79, pp. 1070–1076, 1991.
6. J. Pan, "Present and Future of Synchronization in the US Telephone Network," *IEEE Transactions on Ultrasonics, Ferroelectrics, and Frequency Control*, vol. UFFC-34, no. 6, pp. 629–638, Nov. 1987.
7. J. E. Abate, E. W. Butterline, R. A. Carley, P. Greendyk, A. M. Montenegro, C. D. Near, S. H. Richman, and G. P. Zampetti, "AT&T's New Approach to the Synchronization of Telecommunication Networks," *IEEE Communications Magazine*, pp. 35–45, Apr. 1989.
8. P. Kartaschoff, "Synchronization in Digital Communications Networks," *Proc. of the IEEE*, vol. 79, pp. 906–914, 1991.
9. W. Lewandowski and C. Thomas, "GPS Time Transfer," *Proc. of the IEEE*, vol. 79, pp. 991–1000, July 1991.
10. D. Kirchner, "Two-Way Time Transfer via Communications Satellites," *Proc. of the IEEE*, vol. 79, pp. 983–990, 1991.
11. R. E. Wilson, "Use of Precise Time and Frequency in Power Systems," *Proc. of the IEEE*, vol. 79, pp. 1009–1018, 1991.
12. R. F. C. Vessot, "Applications of Highly Stable Oscillators to Scientific Measurements," *Proc. of the IEEE*, pp. 1040–1053, 1991.
13. J. R. Vig, "Military Applications of High Accuracy Frequency Standards and Clocks," *IEEE Transactions on Ultrasonics, Ferroelectrics, and Frequency Control*, vol. 40, pp. 522–527, 1993.
14. C. Alley, "Relativity and Clocks," *Proc. 33rd Annual Symposium on Frequency Control*, pp. 4–39, 1979.
15. G. M. R. Winkler, "Synchronization and Relativity," *Proc. of the IEEE*, vol. 79, pp. 1029–1039, 1991.
16. J. R. Vig, "Quartz Crystal Resonators and Oscillators for Frequency Control and Timing Applications: A Tutorial," Jan. 2000.
17. "Improving the Accuracy of a Crystal Oscillator," AN1200.07, available: www.semtech.com/images/datasheet/xo_precision_std.pdf, SEMTECH.
18. Samira Zaliasl et al. "A 3 ppm 1.5 × 0.8 mm2 1.0 μA 32.768 kHz MEMS-Based Oscillator," *IEEE J. Solid-State Circuits*, vol. 50, pp. 291–301, Jan. 2015.
19. L. Xiu, "The Concept of Time-Average-Frequency and Mathematical Analysis of Flying-Adder Frequency Synthesis Architecture," *IEEE Circuit and System Magazine*, pp. 27–51, Sept. 2008.
20. L. Xiu, "Nanometer Frequency Synthesis Beyond the Phase-Locked Loop", John Wiley/IEEE Press, Piscataway, NJ, 2012.

21. L. Xiu and P. L. Chen, "A Reconfigurable TAF-DPS Frequency Synthesizer on FPGA Achieving 2 ppb Frequency Granularity and Two-Cycle Switching Speed," *IEEE Trans. on Industrial Electronics*, vol. 64, pp. 1233–1240, Feb. 2017.

22. Atlys™ Board Reference Manual, DIGILENT, Aug. 2013, https://www. digilentinc.com/Data/Products/ATLYS/Atlys_rm_V2.pdf.

23. H. Mair and L. Xiu, "An Architecture of High-Performance Frequency and Phase Synthesis," *IEEE J. Solid-State Circuits*, vol. 35, pp. 835–846, June 2000.

24. L. Xiu and Z. You, "A Flying-Adder Architecture of Frequency and Phase Synthesis with Scalability," *IEEE Trans. on VLSI*, vol.10, pp. 637–649, Oct. 2002.

25. L. Xiu, "A Fast and Power-Area Efficient Accumulator for Flying-Adder Frequency Synthesizer," *IEEE Trans. on Circuit and System I*, vol. 56, pp. 2439–2448, Nov. 2009.

26. L. Xiu, W. T. Lin, and K. Lee, "A Flying-Adder Fractional-Divider Based Integer-N PLL: The 2nd Generation Flying-Adder PLL as Clock Generator for SoC", *IEEE J. Solid-State Circuits*, vol. 48, pp. 441–455, Feb. 2013.

27. W. H. Hsu, K. Y. Tu, J. S. Wu and C. S. Liao, "Frequency Calibration Based on the Adaptive Neural-Fuzzy Inference System," *IEEE Trans. Instrum. Meas*, vol. 58, pp. 1229–1233, Apr. 2009.

28. P. Park, D. Ruffieux, and A. A. Makinwa, "A Thermistor-Based Temperature Sensor for a Real-Time Clock with ± 2 ppm Frequency Stability," *IEEE J. Solid-State Circuits*, vol. 50, pp. 1571–1580, July 2015.

29. P. Chen, C.-C. Chen, C.-C. Tsai, and W.-F. Lu, "A Time-to-Digital Converter-Based CMOS Smart Temperature Sensor," *IEEE J. Solid-State Circuits*, vol. 40, no. 8, pp. 1642–1648, Aug. 2005.

30. P. Chen, C.-C. Chen, Y.-H. Peng, K.-M. Wang, and Y.-S. Wang,"A Time-Domain SAR Smart Temperature Sensor with Curvature Compensation and a 3σ Inaccuracy of $-0.4°C\sim+0.6°C$ over a $0°C$ to $90°C$ Range," *IEEE J. Solid-State Circuits*, vol. 45, no. 3, pp. 600–609, Mar. 2010.

31. P. Chen, M.-C. Shie, Z.-Y. Zheng, Z.-F. Zheng, and C.-Y. Chu, "A Fully Digital Time-Domain Smart Temperature Sensor Realized with 140 FPGA Logic Elements," *IEEE Trans. Circuits Syst. I*, vol. 54, no. 12, pp. 2661–2668, Dec. 2007.

32. P. Chen, S.-C. Chen, Y.-S. Shen, and Y.-J. Peng, "All-Digital Time Domain Smart Temperature Sensor with an Inter-Batch Inaccuracy of $-0.7°C \sim 0.6°C$ After One-Point Calibration," *IEEE Trans. Circuits Sys. I*, vol. 58, no. 5, pp. 913–920, May 2011.

33. T. H. Tran, H. W. Peng, P. Chao, and J. W. Hsieh, "A Low-ppm Digitally Controlled Crystal Oscillator Compensated by a New 0.19-mm^2 Time-Domain Temperature Sensor," *IEEE Sensor Journal*, vol. 17, pp. 51–62, Jan. 2017.

34. A. Ballato and T. Lukaszek, "Higher Order Temperature Coefficients of Frequency of Mass-Loaded Piezoelectric Crystal Plates," *Proc. 29th Ann. Symp. on Frequency Control*, pp. 10–25, 1975.

35. S. Schodowski, "Resonator Self-Temperature-Sensing Using a Dual-Harmonic-Mode Crystal Oscillator," *Proc. 43rd Annual Symposium on Frequency Control*, pp. 2–7, 1989.

36. R. Filler and J. Vig, "Resonators for the Microcomputer-Compensated Crystal Oscillator," *Proc. 43rd Annual Symposium on Frequency Control*, pp. 8–15, 1989.

37. E. Jackson, H. Phillips, and B. E. Rose, "The Micro-Computer Compensated Crystal Oscillator—A Progress Report," *Proc. 1996 IEEE Int'l Frequency Control Symposium*, pp. 687–692.

6

Exploiting Time: The Intersection Point of Multidisciplines and the Next Challenge and Opportunity in the Making of Electronics

Liming Xiu

CONTENTS

6.1 Time: From a Historical Perspective

Along the evolution path of human species, the sense of time originated from the heavenly rhythm of solar bodies. Our sense of flow of time revolves around the sun: a day is defined by the cycle of sunrise and sunset, a month by the cycle of the moon, and a year by the predictable rhythm of the seasons. During a long period, slowly, we built tools to trace the flow of time more predictably: the sundial was used to track the passage of a day, and celestial observation was used to track seasonal milestones, such as the solstice. Gradually, we developed means to measure the flow of time more accurately: dividing time into shorter units of hour, minute, and second. As our tools of measuring time progressively increased in precision, however, some flaws in the celestial metronome began to be noticed: the clockwork of the heavens turned out to be more or less wobbly. Therefore, in 1967, the measurement of time was traded, going from the largest entity in the solar system to one of the smallest in the universe: "the duration of 9,192,631,770 periods of the radiation corresponding to the transition between the two hyperfine levels of the ground state of the cesium-133 atom" (Merriam-Webster). This is the current definition of *second*, and it is the most stable one so far. A "second" is in the structural foundation of the electronic system: flow of time is built from it. It is one of the cornerstones, if not the most important one, of our modern electronic-centric life.

6.2 Five Ws Inside the Electronic System

In all kinds of problem solving, the five Ws—what, who, where, when, and why—are the key means for information processing and decision making. In the case of chip design, and subsequently in the construction of the electronic system, "who" is the electron (collectively represented as voltage or current), and "why" is the motivation for using an electron to represent and manipulate information. The answer to these two Ws was addressed the day Alexander Bell carried out his historical experiment on the telephone in March 10, 1876, and it remains to this day. "Where" is determined by the physical location of a specific piece of circuit within a system. This is controlled by the designer, especially the system-level designer, while a chip project is planned. "What" refers to the activity around the subject of how to use electrons to create an event and then process information. This is the place where a circuit designer uses his or her imagination to create marvels. In this domain, there are two schools of practice: the analog approach, where every voltage point counts, and the digital method, where only two voltage regions, high and low, are meaningful.

The last W, "when," is used to coordinate at what moment an event happens or is scheduled to happen. This W is linked to the flow of time. In our social life, the sense of flow of time is created by the relative movement between the sun and the earth, a water or sand flow, a cyclical movement generated by a mechanical device, or a periodical oscillation (electrical or mechanical). Inside the electronic world, the flow of time is established by a special signal called a clock, which is an electrical pulse train. The key characteristic of a clock signal is its frequency. It is used to gauge the "speed" of an electronic system. Therefore, "when" is mechanically derived from a clock pulse train whose construction is largely dependent on the concept of frequency.

6.3 Two Fundamental Variables: Level and Time

A silicon chip, an electronic system, is used to process information. Information receivable by a human's five senses can show its presence in many different forms: sound, light, pressure, radio wave, mechanical vibration, temperature change, shock from electrical voltage/current, taste from chemical composition, and so forth. When information enters into a silicon chip (electronic system), it can only be exhibited through electron movements, which are collectively represented as electrical voltage and/or current. Voltage and current are then used to create an electrical signal. A signal is subsequently used to create an event. At the end of this chain, the functionality of any silicon chip (electronic system) is realized through events. From the information processing perspective, the electrical signal is the foundational building piece in the construction of a silicon chip and electronic system. It is at the bottom level of abstraction. (The model of the electron/hole is at an even lower level; however, it is usually not directly dealt with by circuit- and system-level engineers but instead by physicists.)

From an operational point of view, when electron movement is used for representing information, voltage is preferred over current in processing simply for convenience (current is usually converted into voltage before being used for information processing). In terms of voltage-based processing, an electrical signal is classified into two types: analog and digital. In the analog approach, the number of electrons involved in a particular task is precisely counted (i.e., every voltage point represents a specific piece of information). In contrast, digital signal processing only differentiates two states: true and false. Thus, two voltage levels of high and low (two clearly distinguishable amounts of electrons) are sufficient.

For describing an electrical signal, the voltage level is only half of the story. The other factor is the indexing. To complete the story, we need to tell what voltage level happens when. Time is the tool for this indexing. It fulfills this role through the clock pulse train. Thus, to define the entire electronic world of magnificence, we need two—and only two—variables: level and time.

6.4 Flow of Time Inside Electronics: From a Clock Pulse Train

Just like the sense of time in our daily life is created by certain kinds of periodic activity (sunrise–sunset, cycle of the moon, rhythm of seasons, cyclical movement of a pendulum, etc.), the flow of time inside the electronic world must be created from periodic action as well. This periodic action is generated from electrical oscillation, which is the alternate change of high and low voltage levels. Electrical oscillation can be created from an electrical

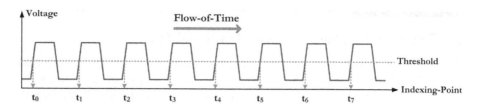

FIGURE 6.1
A sense of time is created through a special electrical pulse train called a clock signal.

circuit without much difficulty. Its output is an electrical pulse train. It is often used in circuit design for achieving many different functions. The special pulse train designed for creating a sense of time is called the clock signal, as shown in Figure 6.1. In operation, a threshold voltage is predefined. During the voltage level's continuous up and down movement, a moment in time is marked whenever the clock signal crosses the threshold. The resulting moments, t_0, t_1, t_2, ..., are the indexing points in time. This creates the movement of time flow (sense of time). Those points can be used for indexing functional events. The efficiency of this clock pulse train is determined by the span between any two adjacent moments (such as t_1 and t_2). The smaller the span, the finer resolution is the indexing. This span is named clock period, and its reverse is the clock frequency. Thus, a high clock frequency can index more events in a given time window. But, on the other hand, it requires more energy to operate.

6.5 Clock Requirements: Precision of Current Practice

Refer back to Figure 6.1; the moments t_0, t_1, t_2, ..., are used as index points to mark functional events. The efficiency of this work depends highly on the accuracy of the locations of those moments. It is preferred that those moments occur exactly at their designed locations. Deviation from the designed value can lead to loss of operating room for functional circuits (digital) or degraded accuracy on signal processing (analog). The degree of this deviation is quantitatively described by jitter (time domain) or phase noise (spectrum domain). Therefore, the key requirement of the clock pulse train is precision: all the resulting moments should be positioned as close as possible to their designed locations.

6.6 Desirable Clock Quality: Flexibility of Future System

Since the concept of the clock pulse train was introduced into electronic system design as the time indexer many decades ago, oftentimes it has been used in the fashion of fixed frequency. In other words, during an operation, the value of the period is constant and the spans between the moments are all equal in length. Implementation-wise, this kind of clock pulse is most suitable for being created with high precision. However, it is rigorous (or rigid). For future complex systems of a dynamic nature (operating environments, data flow, work loading, power consumption, etc.), this rigorous clock is no longer adequate. We want a clock pulse train wherein the moments of t_0, t_1, t_2, ..., can be dynamically adjusted in operation: the span can be created at any length we want (arbitrary frequency generation), and it can be changed from one value to another quickly (instantaneous frequency switching). This new type of clock pulse train with such flexibility is termed flexible clock. It is more suitable for a system wherein clock flexibility has a higher priority than clock precision.

6.7 Making a Flexible Clock: Reinvestigating the Concept of Clock Frequency [1, 2]

The first feature of a flexible clock is arbitrary frequency generation. When the clock pulse train of Figure 6.1 is inspected, it demands that we should have the capability to create a pulse with any desired length in time (controlling the span any way we want). However, unlike voltage level, time is much more difficult to deal with. With today's circuit technology, we can set the voltage level to almost any value we want. But achieving a similar result in frequency (which is derived from time) is much tougher. This is primarily due to the reason that none of the four foundational elements—resistor, capacitor, inductor, and memristor—recognize the concept of time. They can response to electron movement (voltage or current), but they have no way of sensing the flow of time. Therefore, it is fundamentally difficult to control the span between moments (i.e., the frequency of the clock pulse train). For achieving the goal of arbitrary frequency generation, we have to look beyond the locality of each individual pulse; we have to investigate at a bigger-picture level.

In the past practice of electrical engineering, frequency was commonly recognized as the inverse of clock period, that is, the span between the two moments making up a particular pulse. This view implies that all the pulses in a clock pulse train are the same: they have the same length in time. This makes the implementation task easier. This local-oriented view is valid in practice, but it is not the full picture. In a higher view, frequency is defined in a much larger window of 1 second. It is the number of clock pulses that exists in the time window of 1 second when the clock signal is concerned; it is the number of operations executed within the time window of 1 second when functional operation is concerned. This broad view leads to a new possibility: the constraint of "all the cycles must have the same length in time" can be removed. This breakthrough in the clock frequency concept can free our hand

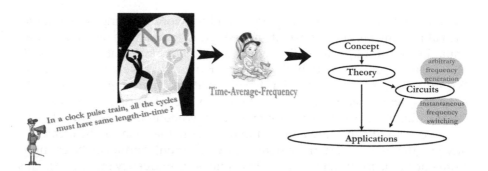

FIGURE 6.2
TAF leads to a new direction and a wave of innovations.

in attacking the problem of arbitrary frequency generation. Traditionally, the clock pulse train is generated inside a chip/system through a special circuit called phase-locked loop (PLL). Due to the feedback mechanism of "compare and then correct," the output frequency from PLL is hard to switch quickly. To make "instantaneous frequency switching" possible, one approach is to directly construct each individual pulse. This is the so-called direct period synthesis (DPS) method.

By removing the constraint of "all the cycles must have the same length in time" and further by directly constructing each pulse, a new type of clock generator, time-average-frequency direct period synthesis (TAF-DPS), emerges. TAF-DPS is the evolution from the rigorous clock to the flexible clock. It is the enabler for innovation in future system, as illustrated in Figure 6.2.

6.8 Opportunities with the Flexible Clock [3, 4, 5]

The essence of all chips and electronic systems is to process information. In chips and systems, at an abstraction level higher than voltage/current, information takes its shape as data flow. Inside the electronic world, the flow of data from place to place is solely controlled by the flow of time: the clock signal. The introduction of a new type of clock, the flexible clock of arbitrary frequency generation and instantaneous frequency switching, provides a new set of tools to circuit designers and system architects. With these tools, as illustrated in Figure 6.3, data flow can be manipulated to a higher degree of freedom. Innovation at a higher level can be made possible.

The steps of actually enabling the innovation wave are illustrated in Figure 6.4. From the breakthrough at the concept level (the novel concept of TAF), a theory is established so that a solid foundation is built. From the concept and theory, circuit development can then be carried out. At this level, four different groups of circuits are explored. The first type is the

FIGURE 6.3
Arbitrary frequency generation and instantaneous frequency switching are circuit-level tools for system-level innovation.

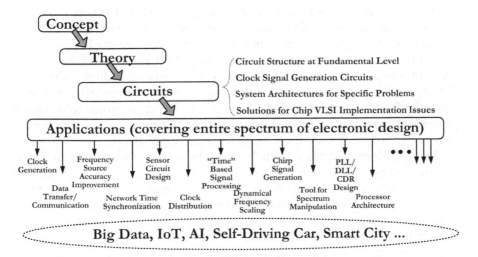

FIGURE 6.4
Steps of enabling an innovation wave from the emerging clock technology.

circuit structure at the fundamental level. It establishes the principle that an electronic system can be driven by a TAF clock (i.e., an electronic system can be driven by the clock pulse train of more than one type of cycle). The second type is a special group of circuits that focus their attention only on generating the clock signal. The third type is system architectures that address specific problems. On this front, the uniqueness of those architectures is that they attack old problems from a new direction: ample frequencies and fast frequency switching. The last type of circuit deals with large-chip implementation problems. As the chip size grows bigger and the process feature size becomes smaller, physical implementation of the complex chip becomes a difficult problem. Clock distribution is one of the toughest issues in this respect. TAF-DPS provides a means of attacking this problem from a new perspective. By applying these techniques to application, these four types of circuits cover almost the entire spectrum of electronic design since everything has something to do with the clock (i.e., the flow of time).

6.9 A Crosspoint of Multidisciplines

Since the invention of the integrated circuit (IC) in the late 1950s, the field of IC design has experienced a continuous evolution. The countless innovations in the areas of microprocessor design, memory structure, and analog/ radiofrequency (RF) technology all focus on using level (i.e., voltage or current level) to accurately and efficiently represent information. In other words, the majority of our circuit work was concentrated on researching "what" (i.e., playing with level). After several decades of brilliant work, it is fair to say that there is not much room left to grow in this direction. However, little attention has been paid to dealing with "when" since the view that *time* serves only the supportive role of indexing is prevalent among IC professionals. With recent advances in process technology, transistors switch faster. This presents a bigger opportunity for us to explore "when." Additionally, from the application perspective, the ever-increasing system complexity demands more flexibility in time indexing. Therefore, to bring electronic information processing capability to the next stage, now is the time to explore the flow of time (when) in greater depth, at a much higher sophistication.

This problem of dealing with "when," however, is nontrivial. It lies at the intersection point among very different disciplines. It is a multidiscipline challenge that requires effort from computer scientists, network architects, VLSI circuit designers, and device physicists, among others. As a professional, working within an established field is both empowering and restricting. On the one hand, staying within the boundary of one's discipline allows him or her to easily make incremental improvements. But on the other hand, the disciplinary boundaries can also serve as blinders, keeping people from bigger ideas that become visible only when those borders are crossed. History has shown many times that ideas travel in networks. An important invention comes into being through the network of collaboration. Once unleashed in the world, a big idea set into motion is rarely confined to a single discipline. For an individual, if he or she wants to improve the world, he or she needs focus and determination. More importantly, it is better for him or her to make new connections than remain comfortably situated in the same routine. For our case of exploiting time and subsequently improving electronic information processing efficiency, the conceptual and technological pieces have already come together to make this campaign imaginable. The purpose of this chapter is to make more people aware of this new direction and inspire them to innovate in this new field.

6.10 Conclusion: Circuits Are Done?

After several decades of effort, the art of designing circuits through semiconductor transistors has reached a very sophisticated level, which leads to the belief that all the important circuit-level problems have already been solved and there is not much room left for circuit designers to make new wonders. In other words, only at higher levels, such as system, architecture, and software levels, can big breakthroughs happen. This belief further leads to the view, among many IC professionals, that circuits are done and excitement can only be found at the higher level. This view has been made popular by recent advances in big data, artificial intelligence, Internet of Things (IoT), self-driving cars, smart cities, and so forth. However, it is worth pointing out that all those high-level activities are built on the foundation of the semiconductor industry. All of them, without single exception, are supported by chips made of transistors. In essence, at the bottom of all those activities lies the operation of controlling data flow (i.e., the stream made of zeroes and ones). Therefore, at the heart of all those activities, they have to deal with the very issue of flow of time, and this handling of time is accomplished through transistors. This fact is graphically illustrated in Figure 6.5. From concept to final product, the deed of creating something useful usually goes through four phases: transistor, geometrical pattern, chip, and system. During this process, the sense of flow of time is established only through the clock signal. In current practice, the clock signal we use is rigorous (or rigid). Consequently, the control of flow of time is rigorous, which results in inefficient handling of data flow. To improve the efficiency of data movement (i.e., the information processing efficiency), we need to make the clock flexible. This influences all aspects of IC design. Flow of time is an eternal issue before, during, and after Moore's law; the better handling of time is therefore

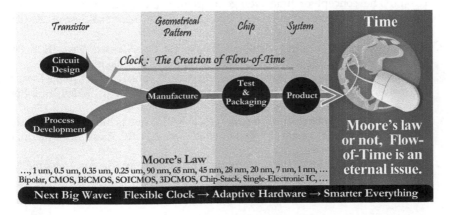

FIGURE 6.5
The clock (flow of time) is the driver for the next big innovation wave.

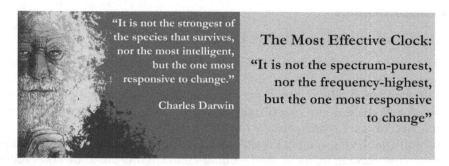

FIGURE 6.6
The most effective clock is the one that is most responsive to change.

expected to exert a profound influence in the decades to come. It is the driver for the next big wave of smarter everything. From this point of view, it is fair to say that circuits are not done yet; they still have lots of room to grow and much excitement remains to be explored. Circuits still have a long way to go, and the direction is to make clocks flexible, and consequently to make everything smarter and every operation more efficient.

As Charles Darwin said more than 100 years ago, "It is not the strongest of the species that survives, nor the most intelligent, but the one most responsive to change." Along this line of thought, the most effective clock is the one that is most responsive to change (Figure 6.6).

References

1. L. Xiu, The concept of time-average-frequency and mathematical analysis of flying-adder frequency synthesis architecture, *IEEE Circuit and System Magazine*, Sept. 2008, pp. 27–51.
2. L. Xiu, *Nanometer Frequency Synthesis Beyond the Phase-Locked Loop*, IEEE Press Series on Microelectronic Systems, Wiley-IEEE Press, Hoboken, NJ, 2012.
3. L. Xiu, *From Frequency to Time-Average-Frequency: A Paradigm Shift in the Design of Electronic System*, IEEE Press Series on Microelectronic Systems, Wiley-IEEE Press, Hoboken, NJ, 2015.
4. L. Xiu, Spectrally pure clock vs. flexible clock: Which one is more efficient in driving future electronic system? in *Mixed-Signal Circuits*, CRC Press, Boca Raton, FL, 2015.
5. L. Xiu, Clock technology: The next frontier, *IEEE Circuit and System Magazine*, May 2017, pp. 27–46.

7

Aging Evaluation and Mitigation Techniques Targeting FPGA Devices

Ioannis Stratakos, Konstantinos Maragos, George Lentaris, Dimitrios Soudris, and Kostas Siozios

CONTENTS

7.1 Introduction

Digital circuit downscaling is driven by a continuous need for integrated solutions that deliver higher performance in the smallest possible size. Field-programmable gate arrays (FPGAs) are an attractive solution for use in the implementation of digital systems, because they exploit the latest fabrication processes on complementary metal-oxide semiconductor (CMOS) technology in order to provide the highest possible performance with power consumption as low as possible. However, the adoption of the latest fabrication process comes with great challenges, such as process variability, power dissipation, and reliability issues due to transient and permanent errors.

Regarding reliability, which plays a crucial role in a digital system, transistor aging is an important cause of permanent failures after a long operation time.

The mechanisms behind aging in digital circuits have been well known for a long time, but a systematic exploration of their effect in FPGA devices started to become relevant quite recently, because of the continuous market demand for more performance in the most compact size. FPGAs are rather interesting platforms for studying aging-induced degradation phenomena, due to their regular architecture, compared with an *application-specific integrated circuit* (ASIC). FPGAs are composed of a small number of basic resources uniformly placed across the chip, along with the appropriate communication infrastructure for connecting these resources together. Some FPGA resources may be unused or partially used after the placement and routing of a system on an FPGA. One other interesting fact about FPGAs that relates to aging is that because the final implemented design is not known a priori, the person or team creating the system may not know the exact location of the critical paths. So aging-aware techniques applied to ASIC designs can not be used directly. So, one must find ways to either adapt ASIC-based techniques to FPGA devices, which in most cases is very difficult or even impossible, or create new tools and techniques tailored to FPGA devices, which must take into consideration their unique features, such as their regular architecture and their reconfiguration capabilities.

The rest of this chapter is organized as follows. In Section 7.2, the main factors that contribute to the aging of electronic circuits, which also apply to FPGA devices, are presented. Then in Section 7.3, an overview of methods for evaluating aging-induced degradation is given, with some examples from the literature. Section 7.4 gives an overview of proposed techniques to mitigate the effects of aging on FPGA-based systems. Finally, Section 7.5 concludes the chapter.

7.2 Aging Mechanisms in Digital Circuits

After an electronic device has been shipped to customers, it is expected to operate correctly under predefined conditions (e.g., voltage, temperature, frequency), determined during the test phase, for its full operation life cycle. However, continuous operation in line with vendor specifications is hard to achieve. Physical changes on the transistor level, such as charge movement and breaking of bonds, are inevitable. This leads to performance degradation over time. This degradation phenomenon, called aging, is one of many reliability issues that very large-scale integration (VLSI) devices face, and it becomes more intense as smaller transistor feature sizes are used.

Aging develops gradually over a long time period, in which the performance of the device decays until a critical point is reached. Before this critical point, the effects of aging in a device start to show up in the form of transient errors that in most cases can be tackled and the correct operation of the system can be maintained. However, when the device passes the critical point, the errors that occur start to become permanent, leading to destructive conditions for the system. The parts that are affected the most from aging are the oxide on the gates of transistors and the interconnection infrastructure of the circuit. As time passes, circuits are subject to stress conditions, which decay the properties of the transistors, leading to faulty operation conditions. The most obvious effect of transistor aging is the increase of the propagation delay, which translates to a lower operating frequency than the one the vendor has defined. The main mechanisms that affect circuit aging are

- Bias-temperature instability (BTI)
- Hot-carrier injection (HCI)
- Time-dependent dielectric breakdown (TDDB)
- Electromigration

The first three aging mechanisms affect the transistors of a circuit, while the last one is the main cause of failures in the interconnection infrastructure of a digital circuit. High temperature plays a significant role in accelerating the aging rate in a digital circuit; hence, the circuits age faster the higher their temperature is. Other factors that enhance the aging effects of the above mechanisms are high switching activity on transistors and high supply voltages [1]. These factors relate to the operational conditions of the system. However, there are other factors that affect the reliability of a system and in a way relate to aging. Specifically, as circuits become more complex, the number of resources that age increases significantly, and so does the number of possible points of failure in a system. Next, a brief overview of the above-mentioned aging mechanisms is given.

7.2.1 Bias-Temperature Instability

Bias-temperature instability is a degradation phenomenon, affecting mainly metal-oxide-semiconductor field-effect transistors (MOSFETs). It manifests as an increase in the absolute value of the threshold voltage (V_{th}) on the transistors. Until some time ago, the highest impact was observed in pMOS transistors when they were reversed biased (negative BTI [NBTI]). With the introduction of fabrication processes, for 32 nm technologies and less, the use of metal gate/high-k CMOS technology gave rise to a similar phenomenon, the positive BTI (PBTI), which affects nMOS transistors, and which until then, its impact was negligible compared with NBTI. The main factor for BTI degradation, either NBTI or PBTI, is dangling bonds, called *traps*, that develop between the channel and the oxide layer when an electric stress is applied across the gate oxide and leads to hydrogen diffusion. BTI manifests itself as an increase in the absolute threshold voltage V_{th} of the transistor over time and includes two phases:

1. *Stress phase*: During this phase, when the transistor operates (ON state), interface traps accumulate between the channel and gate oxide, leading to an increase of the threshold voltage V_{th} and, as a result, degradation in transistors, turn-on delay (see Figure 7.1(a) for NBTI).

2. *Recovery phase*: In this phase, the transistor does not operate (OFF state) and a partial reduction of interface traps takes place; thus, V_{th} decreases, leading to a partial recovery of the original turn-on delay. This phase cannot fully negate the effects of the stress phase, and as a result, the final outcome over a period of time is an increase in transistor threshold voltage.

So, it is evident that the BTI aging mechanism consists of a static and a dynamic part. Figure 7.2 presents the dynamic effect on V_{th} of NBTI degradation over time. As can be seen during the stress phase, an increase of V_{th} is

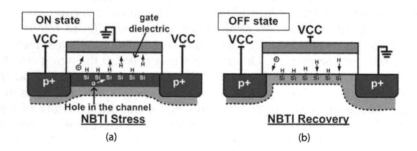

FIGURE 7.1
The two phases of the NBTI aging mechanism. PBTI manifests based on the same principle, but for nMOS transistors [2].

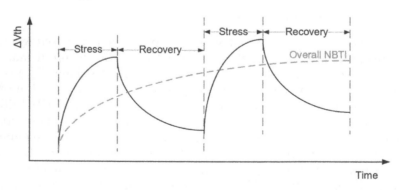

FIGURE 7.2
Dynamic (black continuous line) and static (red dashed line) effect of NBTI in transistors V_{th} [3].

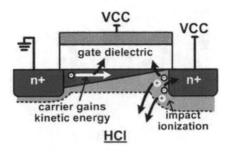

FIGURE 7.3
Physics behind the HCI aging mechanism [2].

observed, while in recovery the threshold voltage decreases, but a full recovery to its original state before the stress is not observed. So, after a long operation time of a circuit, this will translate into absolute increase in V_{th}, as shown by the red dashed line in Figure 7.2.

7.2.2 Hot-Carrier Injection

Hot-carrier injection is another aging mechanism that also affects the performance of a circuit. The cause behind this phenomenon is the accumulation of trapped carriers in the gate oxide, which are created by high-energy carriers able to overcome the potential barrier of the gate dielectric. As time passes, these trapped carriers create a charge that increases V_{th} and decreases the mobility on the area, leading to a slower transistor (Figure 7.3). Unlike BTI, HCI manifests only as a dynamic phenomenon and is heavily influenced by the switching activity of the transistor; thus, its effect on the transistor can not be compensated at all. In a digital circuit, an area with low switching activity is affected less than another with a higher one.

7.2.3 Time-Dependent Dielectric Breakdown

As the thickness of gate oxide is reduced, *time-dependent dielectric breakdown* becomes one of the main causes of permanent failures in today's digital circuits. TDDB is caused by the creation of a conductive path passing through the gate oxide. This path is the result of defects or trapped charges due to high gate voltage in a transistor (see Figure 7.4). A leakage current flows through this conductive path and results in increasing the switching delay of the transistor, and in extreme conditions, the transistor may not switch at all.

7.2.4 Electromigration

Electromigration is the main cause of failures in the interconnect infrastructure of a digital circuit, which can lead to permanent errors. The reason behind

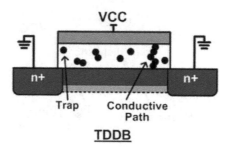

FIGURE 7.4
Physics behind the TDDB aging mechanism [2].

electromigration is the migration of metal ions over time. High-density currents flowing through the circuit wires are the main reason for this migration. The immediate consequence of electromigration is an increase of wire impedance, and as time progresses, the final outcome is the development of open and short circuits. Moreover, the smaller the feature size of the wires, in a technology node, the more intense the electromigration effect becomes.

7.3 Evaluation of Aging in FPGAs

There are a plethora of proposed ways to assess the performance degradation due to aging in digital circuits that can be applied to FPGA devices as well. Two widely used approaches are

1. The offline methods that mainly use models, for either the FPGA resources or the aging mechanisms. This approach is based on simulating and to a degree predicting the aging-induced effects on a circuit. These kinds of methods use tools such as SPICE, and sometimes information from vendor tools (e.g., from static timing, placement, and routing reports) is used in combination with SPICE models in order to provide more accurate results.

2. The online methods that use measurement circuits implemented in the FPGA fabric to perform real-time estimation of aging in a device. Moreover, applying accelerating aging conditions, such as elevated temperature and operation voltage, is a common procedure used during online aging estimation.

Additionally, there is the possibility of using a combination of the above methods. Next, examples of the two mentioned methods are presented, together with estimated aging degradation.

7.3.1 Offline Evaluation of Aging

Because the exact implementation of an FPGA device is proprietary information and vendors are reluctant to provide information other than what they think is safe to disclose, the offline evaluation of aging most of the time is based on a number of assumptions that must be applied by the user mainly for the underlying architecture of the FPGA (e.g., lookup table (LUT) implementation). Except for the models used for various FPGA resources, there must also be some model for the degradation effect of aging mechanisms on key parameters of resources (e.g., transistor switching delay). A model used quite often in literature for the increase in switching delay on pMOS transistors caused by NBTI is [4]

$$\Delta d = A_{BTI} \times Y^n \times t^n \times e^{\frac{-E_a}{\kappa T}} \times d_0. \tag{7.1}$$

The parameter A_{BTI} is related to the technology used, t is the operation time (age) of the transistor, Y is the duty cycle, T is the temperature, κ is Boltzmann's constant, E_a a fitting parameter, n depends on the fabrication process used, and d_0 is the original switching delay (before aging). Because PBTI is a similar phenomenon, the same equation can be used to model the switching delay degradation on nMOS transistors. A similar model exists for the HCI degradation mechanism, which affects nMOS transistors. HCI switching delay degradation of a transistor can be modeled as [4]

$$\Delta d = A_{HCI} \times \alpha \times f \times t^{0.5} \times e^{\frac{-E_b}{\kappa T}} \times d_0. \tag{7.2}$$

As previously, A_{HCI} is related to the technology used, t is again operation time (age) of the transistor, E_b is a fitting parameter, α is the activity factor, f is the frequency, and d_0 is the original preaging switching delay. As can be seen from the previous two models, BTI and HCI strongly depend on fabrication process, temperature, and device age. All play significant a role in aging, so in order to be able to evaluate aging through models, a great deal of effort must be made to find the appropriate values for the different parameters.

In [5], the effect of NBTI and PBTI aging is investigated for a number of different two-input LUT implementations using SPICE simulation for the LUT architecture, as well as the voltage threshold shift due to NBTI and PBTI aging. The authors investigated the aging effects on three common LUT architectures:

1. Logic-gate-based structure
2. Pass-transistor-based structure
3. Transmission-gate-based structure

The test procedure consisted of two experiments. The first experiment considered only the NBTI, which was the main BTI aging factor in older FPGA technologies, while the second took into consideration the combination of NBTI and PBTI. For the aged LUTs, the underlying transistors were replaced with the equivalent SPICE models that take into consideration the voltage threshold shift, and the targeted technology was based on the 22 nm predictive technology model (PTM) for metal gate/high-k CMOS. The results show that not all LUT architectures are suitable when metal gate/high-k CMOS technology is used, which is the case in most modern FPGA devices. Specifically, the pass-transistor architecture is shown to be the least optimal choice, even though it has minimum area. Moreover, the authors concluded that the input signal probabilities directly affect the aging rate of the transistors, and that the LUT configuration has a considerable impact also. Another interesting conclusion is that the default all-zero configuration for unused LUTs, which major FPGA vendors have adopted, is not the most optimal one regarding aging rate, as it may lead to higher degradation. However, using information obtained through simulation, with knowledge of the LUT structure and its current configuration, a better default configuration can be found.

The routing resources are another building block that is abundant in FPGA devices. That is why an investigation for the degradation effect on these must also be performed. Because routing resources are also constructed using transistors, the same degradation mechanisms apply (BTI, HCI, etc.). However, this time other parameters must be considered. First, routing resources are connected together in a convenient manner to support signal transmissions throughout the FPGA chip. Second, many of the routing resources will derive multiple destination points from a single source point. Also, the way the routing resources are connected together will have an impact on their degradation rate. The effects of all these constraints are evident in [3], where again, based on simulations using SPICE, the combined effect of NBTI and PBTI is studied on different architectures of routing FPGA resources in relation to wire length, number of serial connected switches, and their fan-out. The models used are based on a 32 nm PTM library and four routing resource structures are considered:

1. Pass-transistor with keeper
2. Tri-state buffer
3. Transmission gate
4. Multiplexer

Moreover, for the wires an RC model was used. The authors concluded that by using a routing infrastructure that utilizes a small number of cascaded switches, together with long wires, the aging degradation is less compared with that when using a larger number of switches with short wires. In addition, high fan-out for the tristate buffer architecture, as well as higher values of V_{dd}, leads to lower aging degradation. The same relation to V_{dd} also seems to apply to the pass transistor with keeper architecture. Their last conclusion suggests that based on the technology used, the architectures based on nMOS transistors are the best candidates when the NBTI effect is the dominant factor of aging-induced degradation, while the transmission gate architecture is best suited when the effects of NBTI and PBTI are considered.

Moving to the application level, the work in [4] investigates how a design mapped to an FPGA affects its aging degradation. In order to be able to have the most accurate results, the authors extracted information from commercial tools regarding the timing of paths of the design, the activity rates, and the signal probabilities. Together with a thermal profile created from HotSpot, they performed simulations using SPICE and estimated the effects of aging on the implemented design. The device used was a 40 nm Virtex-6 FPGA from Xilinx. From the experiments they concluded that the implemented design has a direct impact on the aging of the FPGA device and, as a result, its lifetime and reliability. Moreover, the way the same design is mapped to the device will affect the aging rate differently regardless of the optimization algorithm used.

In [6], the NBTI effect, not only on combinational circuits but also on sequential circuits, is the main topic of interest. Even though this work is not FPGA specific, its conclusions can be adopted, because FPGAs consist of a sea of combinational (LUTs, muxes, and carry chains) and sequential (flip-flops) digital components. The study is based on a developed framework and SPICE tool, for a 65 nm, technology node, and correlates supply voltage, temperature, switching activity, and input pattern to NBTI degradation. For the combinational circuits, the results show that for the most optimal input pattern, when the circuit is idle, the performance degradation can be up to ~12% compared with the worst-case input pattern (degradation close to 50%). Regarding supply voltage and temperature, the conclusions are that temperature degrades the performance of a circuit much more than the supply voltage. Moreover, temperature variation can impose a 3.5% difference in degradation, while voltage variations impose only a 0.3% difference. Regarding the clock distribution network, the results suggest that NBTI has a trivial effect; thus, the rise and fall edges of the clock signal are not affected by aging, so is the frequency of the clock. That behavior is mostly related to the fact that the clock signal is produced from an oscillator external to the system. This last observation is quite relevant to FPGAs, as the operating clock of the device comes from external sources and all the other clocks used in an FPGA-based system are generated from this main clock in specialized resources inside the FPGA fabric.

7.3.2 Online Evaluation of Aging

The use of online methods to measure aging degradation on FPGA devices is the most common mechanism for real-time monitoring. The main idea of online sensing is to use additional circuits that will be able to measure the effects of aging on the FPGA device. The methods used most often for online monitoring are

1. A sensing network based on ring oscillators (ROs). Using the RO network, the user is able to measure any kind of variation the device is experiencing, including aging-induced delay degradation.
2. Extra (shadow) registers bound to the most critical or near-critical registers of the system. Both registers, the original and its counterpart, essentially latch the same data signal, but the way they operate is not the same. There are two possible options. The first one is to operate the extra registers with a variable-phase-offset clock signal, and the second is to use the same clock signal but to add some extra delay to the data signal before reaching the shadow register. In either case, a comparison of the outputs will inform if a violation in the form of late transition occurred. The late transition, if it happens, is an indication of aging degradation.

A key consideration when using online sensors to evaluate the aging of an FPGA in real time is the induced resource overhead of implementing the sensors, due to the extra logic utilized. Moreover, for the RO-based sensing circuit a timing schedule must be defined. This schedule must interleave the useful operation of the system with the measurement procedure in the most optimal way, so that the overhead of monitoring the FPGA does not interfere with the actual operation. Regarding the shadow register approach, compared with the original implemented system, the augmented one with the sensors will have an alternate mapping and routing, thus changing the original paths of the system to be monitored. For these reasons, extra design and implementation constraints, related to mapping, placement, and routing, must be applied to the final system as well as to the sensor infrastructure.

In [8], a critical path monitoring infrastructure is presented. The proposed sensor, shown in the center of Figure 7.5(a), is placed in parallel with the path under investigation, and its signal is used as the clock source for two flip-flops, while the actual clock drives the input pin of the flip-flops. By XORing the output signals of the two flip-flops, late signal transitions can be flagged. The authors also discuss other parameters that can affect sensor sensitivity, such as the detection window. Moreover, a selection scheme is presented to select the most appropriate paths for monitoring. By doing simulations on 40 nm Virtex-6 FPGAs, as well as emulation on an actual FPGA device (65 nm Virtex-5 based XUP-5 board), the feasibility of the sensor becomes evident. The proposed sensor imposes a $\sim$1.3% area overhead, $\sim$1.6% performance degradation, and $\sim$1.5% increase in power consumption. However, the proposed sensor and design flow are not application independent.

On the other hand, in [9] a more general sensor circuit, shown in the left of Figure 7.5(a), and sensor insertion policy are presented. The extra resources needed to implement a sensor are one LUT and four flip-flops per sensor and

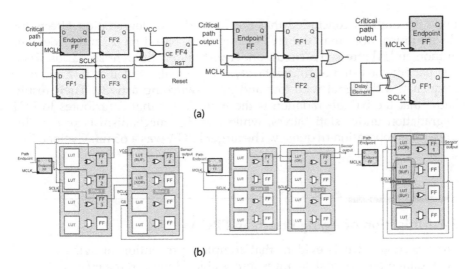

FIGURE 7.5
Different FPGA aging sensors in the literature [7]. (a) Sensor architecture and (b) FPGA resource utilization.

one digital clock management (DCM), which is used globally. The proposed methodology to find candidate paths uses reports produced from vendor tools, and insertion of the sensor can be included in the standard design flow for an FPGA-based system. The sensor was tested on a 90 nm Spartan-3AN FPGA implementing different International Symposium on Circuits and Systems (ISCAS) benchmarks in preaging conditions, as well as during accelerated aging conditions (increased voltage supply and temperature). The most interesting result is that the functionality of the DCM was hardly influenced by the accelerated aging conditions, which suggests that the sensor is robust enough. Moreover, the response of the sensor showed a linear trend even under aging conditions.

In [7], another sensor infrastructure, shown on the right of Figure 7.5(a), is presented and compared with the ones mentioned previously [8, 9]. This sensor occupies less area than the others, and the power overhead is also smaller. Also, the performance degradation due to extra logic is less than ~1% when 15 sensors are used. The key feature of this sensor is that, under emulated aging conditions, it was able to flag error conditions due to aging much sooner than the other two sensors. This makes the proposed sensor architecture much more sensitive. Here it must be noted that the test was performed on a 28 nm Artix-7 FPGA.

Finally, in [10] a number of RO-based sensing circuits with different lengths were used to evaluate how switching activities and signal probabilities affect the aging of an FPGA device. Using a 45 nm Spartan-6 FPGA and stressing the device under accelerated aging conditions for 1 week, ~6% performance degradation was observed. Moreover, for the specific technology node the authors concluded that the BTI-induced degradation affects the device much more than the HCI degradation. Based on a similar idea, the authors in [11] stressed a circuit-under-test under different scenarios. Their test pattern consisted of constant duty cycle signals as well as low and high frequency with signal with low and high switching activity. Their results show that the BTI degradation is the main factor that contributes to LUT degradation under static stress, while the HCI mechanism is responsible under high operation frequency. The target FPGA was a 65 nm Cyclone III.

7.4 Mitigation of Aging Effects in FPGAs

From Section 7.2 it is evident that complete prevention of aging-induced degradation on a digital circuit is impossible. However, the effects of aging can be reduced, in order to have an increased lifetime and reliability of an FPGA-based system, by applying a diverse number of techniques targeting the different aging mechanisms in these devices. For FPGAs, which are by design reconfigurable devices, and most of the time resources available on a device are not fully utilized, reconfiguration schemes seem to be the straight-forward method to adopt. The main idea behind the use of reconfiguration in order to mitigate aging effects is the possibility to evenly distribute the stress applied in FPGA resources. In real-world systems, even resources that are not used in the implemented design have a default configuration that affects the degradation rate of these resources. By applying an new properly chosen FPGA configuration, that does not compromise systems functionality, the aging rate can be reduced and the system can be reliable for an extended period of time. In addition to reconfiguration schemes, researchers have rec-ognized that the internal operation state of the FPGA resources (e.g., signal and node activity) has a significant impact on the aging rate, so for a complete aging mitigation approach, these parameters must be taken into account.

As was the case with evaluating the aging effects on FPGAs, the major-ity of the methods proposed to mitigate aging effects can be categorized mainly into offline methods and a combination of offline methods with a runtime reconfiguration of the FPGAs. Moreover, some of these methods utilize the models presented in Section 7.3.1. In addition to these models, many approaches for aging mitigation extract information from vendor tools. These are mainly related to timing information of critical paths in a cir-cuit and on the placement of a system in the resources of the FPGA, as well as their connection. The main idea of the offline methods is the uti-lization of aging-aware computer-aided design (CAD) tools, which can be executed during the initial design phase of an FPGA-based system, and after the deployment of the system the configuration stays the same. The other methods are based on periodic degradation measurements taken during the operational life of a system in order to adapt the current configuration of the underlying hardware, so that the reconfiguration capabilities of an FPGA are fully utilized. In these kind of methods, the mitigation strategies applied to the system can be agreed on during the design phase in combination with information about the operational environment and conditions. Next, an overview of methods for aging mitigation is given, along with some experimental results from the literature to support the benefits obtained.

7.4.1 Offline Aging Mitigation Techniques

The main characteristic of the offline approach is the use of CAD tools, which are used not only for the implementation of the developed system, but also for the application of the appropriate aging mitigation techniques. This approach mainly covers attempts made during the design phase in order to enhance the overall reliability of a system and its robustness against aging degradation. In addition to the normal design decisions during the initial design phase (e.g., regarding power consumption), models for aging degradation mechanisms must be integrated into the internal timing models used. For commercial tools, the integration of aging models is not straightforward due to the fact that they are proprietary and interventions are rarely easy to apply. However, if the tool provides some way of creating design checkpoints, then these can be used as starting points for invoking the aging mitigation methods externally to the tool. On the other hand, academic CAD tools are more versatile. The designer has all the freedom he or she needs to integrate aging models as well as include mitigation techniques in the regular design flow and generally to customize the tool to specific requirements of the system under development.

Using an offline method for aging mitigation has the advantage of creating a single configuration file and no further actions must be taken after system deployment. Moreover, the end user is completely oblivious regarding the aging mitigation additions to the tool, and due to the fact that the implementation tools are used during the normal design flow of a system, the correct timing is guaranteed. On the other hand, this early planning for aging mitigation has its downsides. Specifically, the designer may not know exactly the input data pattern to the system that can lead to wrong assumptions regarding the aging degradation mechanisms that are going to play a key role in systems deterioration. Also, preemptive measures for a wide range of usage scenarios may render diminishing aging mitigation benefits.

In [12], the BTI aging-induced degradation due to LUT configuration and the input signal probabilities is explored. A tool, developed by the authors, performs an exhaustive search of the best alternative LUT configuration that has a minimum aging effect. The search is conducted based on two choices. The first one is to allow only the LUT configuration to change but not the routing of the circuit, while second one allows rerouting of the circuit. They validate their methodology by modifying the verilog-to-routing (VTR) academic tool and using a device model based on a 40 nm Stratix IV FPGA from Altera. They concluded that the approach that also uses rerouting offers better aging mitigation capabilities, which for BTI is around 20%, and the lifetime of the FPGA device can be improved by around 200%.

In [13], an investigation of the aging impact of HCI, TDDB, and electromigration on the operational lifetime of an FPGA device was conducted. The authors propose a method for each one of them in order to increase the mean time to failure (MTTF) of the FPGA. For HCI mitigation, periodic remapping of the system to less stressed areas of the chip is proposed, resulting in around a 28% increase in lifetime but with about 1.53% performance degradation. To combat TDDB, they employ a method, proposed in [14], to minimize current leakage and report 24% improved MTTF, on average, for different benchmark circuits. Also, in order to reduce the effect of electromigration, a dynamic reconfiguration scheme is used to reroute the connection between resources until an optimal solution is found. The proposed reroute scheme offers a 14.1% life increase with 1.33% frequency reduction, on average.

In addition to the previous works, in [15] the routing resources are of interest. The authors modified the conventional routing algorithm of VTR in order to enhanced it with aging-aware capabilities. Using an island-style FPGA architecture, they demonstrate a ~18% reduction in the aging rate degradation. They claim that this is equivalent to a ~130% improvement in the lifetime of the FPGA device.

From a more general perspective, which is not relate directly to FPGAs, the work in [6] suggests that by applying node activity optimizations, a two to four times reduction in performance degradation, caused by NBTI, can be achieved. Also, by lowering the temperature of a chip, a 60% reduction on NBTI degradation can be achieved.

7.4.2 Aging Mitigation during Runtime

The main idea behind runtime aging mitigation is to adapt the system based on measurements of degradation of the FPGA devices. These methods utilize the online sensors presented in Section 7.3.2, namely, the shadow register or a network of ROs. These sensors are utilized in order to have a real indication of the condition the FPGA device is in. This approach is very versatile because the system can adapt to its current state with a new configuration and prevent possible failures due to aging degradation. Another advantage, compared with offline methods, is that the degradation results obtained are free of possible inaccuracies of the models used, either for aging mechanisms or FPGA resources. A key concept of these methods is that the measurement of degradation due to aging can give information about the aging mechanism that currently has the most severe effect on the device. Based on this knowledge, the aging mitigation strategy used can a target-specific degradation mechanism.

The idea behind this method is that the stress applied to FPGA resources, which leads to aging-induced degradation, can be uniformly distributed to all the available resources of the FPGA. These methods are based on a two-step methodology. The first step consists of creating a number of configuration files for the implemented system, each with alternative mapping, placement, and/or routing, but without compromising the original functionality of the system to be implemented. The second step is the creation of a reconfiguration schedule and its support software and hardware infrastructure, which will be deployed during the operation of the system and will reconfigure the device in predefined timestamps that have been decided. The time when the reconfiguration will be performed can be periodic or not.

From the reasons presented so far, it is evident that aging mitigation methods during systems runtime are an attractive solution because they have the capability to take advantage of the maximum performance; even an aged device can offer and still provide reliable operation. However, there are some drawbacks that are mainly related to the sensing infrastructure, because the sensing logic adds considerable hardware overhead to the system. Additionally, external hardware (e.g., processor and memory) may be deployed in order to perform the measurements and execute the implementation flow to create a new configuration. Additionally, the communication of the external monitoring hardware with the FPGA device adds extra overhead.

The runtime aging mitigation approach is adopted in [16], where an aging-aware floorplanner is introduced. This work takes into consideration aging degradation that relates to BTI and HCI aging mechanisms and presents a framework that, given the proper combination of inputs and constraints, creates a number of physical circuit placements, for the target application, together with a reconfiguration schedule. The authors introduce the *aged-delay map* of region R, where the target application will be mapped. This map describes the aging history of nodes inside the region. Using information from vendor tools that have to do with switching activity, timing of the application, and power consumption, and based on a user-defined maximum acceptable aging degradation, new configurations are created. All the configurations that are created are stored in the configuration memory of the FPGA, and based on timestamps they are loaded on the device during operation of the system. The objective of this method is not only to reduce the aging-induced delay degradation in order to increase the lifetime of a system, but also to create configurations that are able to reduce the aging rate of the device. The authors implemented three real-life applications (AES, DCT, and JPG) and reported up to a 53.2% mitigation on the aging rate of resources and up to a 17.5% reduction in critical path delay degradation. The target device used for the experiments was a 40 nm Virtex-6 FPGA from Xilinx, and it was assumed to be in continuous operation for 3 years.

In [17], a LUT relaxation scheme is proposed. The main idea is to load an inverted bit configuration during the lifetime of the FPGA device. Also, it is noted that the configuration of the interconnection, with the relaxed friendly configuration, is challenging cause of the possibility of routing disruption between CLBs, which in the end change the correct functionality of the system. By proposing a method to change the configuration bits of interconnection switches and applying the reroute method in [18], a decrease of 2.5% on *failure in time* (FIT) is observed. Moreover, a validation of the method in some reference designs from Xilinx managed to recover ~52% of the degraded *signal-noise margin* (SNM).

Another work that tries to overcome the effects of degradation due to BTI mechanisms is [19]. In this work, a three-step exploration framework is presented. The exploration is implemented after the postsynthesis step, conducted by a commercial tool, and its goal is to find an alternative flipped configuration of the original configuration of LUT resources of the FPGA. The original configuration and the flipped one are periodically swapped on the FPGA device. In experiments conducted on a 40 nm Virtex-6 FPGA from Xilinx, on average a 70% improvement was observed in the reduction of SNM and an increase in *soft error rate* (SER).

7.5 Summary

This chapter presented some aging-related facts and research targeting FPGA devices. Due to adoption of the latest process technologies for FPGA fabrication, these devices are most susceptible to phenomena that affect their well-being. Thus, robust methods must be applied to protect FPGAs from aging effects, in order to be able to provide more reliable systems.

Because node downscaling enhances the effects of aging mechanisms (BTI, HCI, TDDB, and electromigration), several works have been presented that study how to evaluate the aging-induced degradation of an FPGA-based system. Investigating the aging effects for the target digital system and technology node offline can provide useful information about possible factors that can reduce a system's reliability. Provided that the models used for FPGA resources and the aging degradation mechanisms are accurate, the results obtained through the offline simulation will be quite accurate themselves. So, the offline methods can be used as guidance in order to decide on possible proactive measures against aging, and as a result shield the correct functionality and reliability of an FPGA device. On the other hand, online methods can be used for real-time estimation of aging. The online methods are based on application of appropriate logic to the device, which act as sensors able to measure aging degradation during the operation of a system, so that appropriate actions can be applied in order to reduce the aging rate of the FPGA device and extend the lifetime and reliability of the implemented system. Table 7.1 presents the main factors that affect the different aging mechanisms. As can be seen, operation temperature and voltage contribute to all

TABLE 7.1
Factors That Affect Aging Degradation Mechanisms

Aging Mechanism	Temperature	Voltage	Duty Cycle	Switching Activity
BTI				
HCI				
TDDB				
Electromigration				

TABLE 7.2
Aging Evaluation Methods

Evaluation Method	Technique
Online	Ring oscillator Shadow register
Offline	SPICE, custom tool Aging mechanism models FPGA resource models

TABLE 7.3
Various Mitigation Techniques for Each Aging Degradation Mechanism and the Tools Used

Aging Mechanism	Mitigation Technique	CAD Tools
BTI	Configuration bit inversion Remapping Node activity optimization Lower temperature	VTR VPR
HCI	Remapping	Vendor
TDDB	Current leakage optimization	Custom
Electromigration	Rerouting	

the aging degradation mechanisms. Table 7.2 gives the main techniques used in the literature to evaluate the aging degradation of FPGA devices.

In cases where the FPGA device operates under extreme conditions, the aging degradation can lead to runtime failures and actions must be taken. However, before even trying to find solutions for tackling runtime errors due to aging-induced degradation, maybe the most optimal approach is to try to reduce aging degradation that leads to failures and, as a result, increase the operation lifetime of a system. For FPGA devices, which are by design versatile processing platforms, reconfiguration schemes to mitigate aging effects and extend the lifetime of a system are logical to use. These approaches are based mainly on offline evaluations of aging. Using information provided from aging evaluations frameworks, implementing a scheduler that will remap the system to spare FPGA resources is a straightforward solution. The remapping can be fine- or coarse-grain. In fine-grain, the scheduler remaps the implemented logic by FPGA resources, such as LUTs, to less stressed ones, for which aging effects are negligible; when a coarse-grain approach is used, complete functional blocks (e.g., IP cores) are moved to less stressed areas of the FPGA device. Finally, in addition to reconfiguration policies for aging mitigation, one can employ special configurations and specific input signals to unused resources in order to incur less stress in these resources, and, as a result, age them less. Later on, these "managed" FPGA resources can be used as targets to move the design from the stressed area of the FPGA to the less stressed one, as mentioned previously. Table 7.3 gives various aging mitigation techniques that target specific mechanisms, as well as some possible tools utilized to deploy aging-aware FPGA configurations.

References

1. Maiti, A., Schaumont, P.: The impact of aging on a physical unclonable function. *IEEE Transactions on Very Large Scale Integration (VLSI) Systems* **22**(9) (2014) 1854–1864.
2. Keane, J., Kim, T.H., Wang, X., Kim, C.H.: On-chip reliability monitors for measuring circuit degradation. *Microelectronics Reliability* **50**(8) (2010) 1039–1053.
3. Amouri, A., Kiamehr, S., Tahoori, M.: Investigation of aging effects in different implementations and structures of programmable routing resources of fpgas. In: *Field-Programmable Technology (FPT), 2012 International Conference on, IEEE* (2012) 215–219.
4. Amouri, A., Tahoori, M.: High-level aging estimation for fpga-mapped designs. In: *Field Programmable Logic and Applications (FPL), 2012 International Conference on, IEEE* (2012) 284–291.
5. Kiamehr, S., Amouri, A., Tahoori, M.B.: Investigation of nbti and pbti induced aging in different lut implementations. In: *Field-Programmable Technology (FPT), 2011 International Conference on, IEEE* (2011) 1–8.
6. Wang, W., Yang, S., Bhardwaj, S., Vattikonda, R., Vrudhula, S., Liu, F., Cao, Y.: The impact of nbti on the performance of combinational and sequential circuits. In: *Proceedings of the 44th Annual Design Automation Conference, ACM* (2007) 364–369.
7. Ghaderi, Z., Ebrahimi, M., Navabi, Z., Bozorgzadeh, E., Bagherzadeh, N.: Sensible: a highly scalable sensor design for path-based age monitoring in fpgas. *IEEE Transactions on Computers* **66**(5) (2017) 919–926.
8. Amouri, A., Tahoori, M.: A low-cost sensor for aging and late transitions detection in modern fpgas. In: *Field Programmable Logic and Applications (FPL), 2011 International Conference on, IEEE* (2011) 329–335.
9. Valdes-Pena, M.D., Freijedo, J.F., Rodriguez, M.J.M., Rodriguez-Andina, J.J., Semiao, J., Teixeira, I.M.C., Teixeira, J.P.C., Vargas, F.: Design and validation of configurable online aging sensors in nanometer-scale fpgas. *IEEE Transactions on Nanotechnology* **12**(4) (2013) 508–517.

10. Amouri, A., Bruguier, F., Kiamehr, S., Benoit, P., Torres, L., Tahoori, M.: Aging effects in fpgas: an experimental analysis. In: *Field Programmable Logic and Applications (FPL), 2014 International Conference on, IEEE* (2014) 1–4.
11. Naouss, M., Marc, F.: Design and implementation of a low cost test bench to assess the reliability of fpga. *Microelectronics Reliability* **55**(9) (2015) 1341–1345.
12. Rao, P.M., Amouri, A., Kiamehr, S., Tahoori, M.B.: Altering lut configuration for wear-out mitigation of fpga-mapped designs. In: *Field Programmable Logic and Applications (FPL), 2013 International Conference on, IEEE* (2013) 1–8.
13. Srinivasan, S., Mangalagiri, P., Xie, Y., Vijaykrishnan, N., Sarpatwari, K.: Flaw: fpga lifetime awareness. In: *Proceedings of the 43rd Annual Design Automation Conference, ACM* (2006) 630–635.
14. Anderson, J.H., Najm, F.N.: Active leakage power optimization for fpgas. *IEEE Transactions on Computer-Aided Design of Integrated Circuits and Systems* **25**(3) (2006) 423–437.
15. Khaleghi, B., Omidi, B., Amrouch, H., Henkel, J., Asadi, H.: Stress-aware routing to mitigate aging effects in sram-based fpgas. In: *Field Programmable Logic and Applications (FPL), 2016 International Conference on, IEEE* (2016) 1–8.
16. Ghaderi, Z., Bozorgzadeh, E.: Aging-aware high-level physical planning for reconfigurable systems. In: *Design Automation Conference (ASP-DAC), Asia and South Pacific, IEEE* (2016) 631–636.
17. Ramakrishnan, K., Suresh, S., Vijaykrishnan, N., Irwin, M.J.: Impact of nbti on fpgas. In: *VLSI Design, 2007. Held jointly with 6th International Conference on Embedded Systems, 20th International Conference on, IEEE* (2007) 717–722.
18. Stott, E.A., Wong, J.S., Sedcole, P., Cheung, P.Y.: Degradation in fpgas: measurement and modelling. In: *Proceedings of the 18th Annual ACM/SIGDA International Symposium on Field Programmable Gate Arrays, ACM* (2010) 229–238.
19. Ghaderi, Z., Bagherzadeh, N., Albaqsami, A.: Stable: stress-aware boolean matching to mitigate bti-induced snm reduction in sram-based fpgas. *IEEE Transactions on Computers* **67**(1) (2017) 102–114.

Index